KANBAN CHANGE LEADERSHIP

KANBAN CHANGE LEADERSHIP

Creating a Culture of Continuous Improvement

KLAUS LEOPOLD
SIEGFRIED KALTENECKER

Published by John Wiley & Sons, Inc., Hoboken, New Jersey
Published simultaneously in Canada

Library of Congress Cataloging-in-Publication Data:

Leopold, Klaus.
Kanban change leadership : creating a cluture of continuous improvement / Klaus Leopold, Siegfried Kaltenecker.
pages cm
Includes bibliographical references and index.
ISBN 978-1-119-01970-1 (hardback)
1. Just-in-time systems. 2. Continuous improvement process. 3. Transformational leadership. 4. Job enrichment. I. Kaltenecker, Siegfried. II. Title.
TS157.4.L36 2015
658.5′1–dc23

2014043443

1 2015

CONTENTS

APPRAISALS

Kanban Change Leadership spells out not only what Kanban is but why and how it works in high-variation environments. There is deep thinking and real implementation in this impressive book. *Jim Benson—creator of Personal Kanban and CEO of Modus Cooperandi, USA.*

The authors of the book *Kanban Change Leadership* perfectly mastered the balance between writing the standard and giving the reader a hands-on guideline on how to achieve real change using simple techniques that have an extremely deep impact. Even after reading it multiple times, this book stays exciting and should be—next to David J. Andersons' work—prominently present in every bookshelf. *Eric-Jan Kaak—CIO Tecnica Group SpA/Blizzard Sport, Austria.*

I've been lucky enough to follow Klaus' and Sigi's live presentations closely for years. Now with *Kanban Change Leadership*, everyone can have access to their insightful and practical advice regarding implementing Kanban. Change agents will be especially interested in Klaus' and Sigi's focus on sociological and cultural concerns (as the excellent bibliography clearly illustrates). This is highly recommended for team leaders, coaches, managers, and executives. *Jabe Bloom, Chief Flow Officer, PraxisFlow/Principle Consultant, Coherent Insight, USA.*

Finally a book for practitioners that isn't scared of presenting useful theory as well as concrete tips. Through the entertaining use of storytelling, and drawing from their experience, they do an excellent job of tackling the more difficult topics of change and leadership. *Kurt Häusler, Software Development Manager and Kanban Coaching Professional, Germany.*

Klaus and Sigi provide a comprehensive yet approachable guide to using Kanban to manage change. Starting with a good practical coverage of the principles and practical of the Kanban method from the perspective of how to actually start using it, the book then takes flight into deeper but still practical change management models and practices and then ties it all together in the third part leveraging lifelike stories of change to bring together and liven up all elements in an experiential manner. As an enterprise kanban coach, I found the book inspired me to aim higher, and I will also recommend it to others looking to manage change in evolutionary respectful ways such as Kanban in their organization or as a professional practice. *Yuval Yeret—Enterprise Lean/Agile Coach & CTO AgileSparks, Israel.*

The unique strength of this book is combining the necessary background (WHY?) on change with the necessary WHAT and HOW (Kanban). *Markus Andrezak, CEO überproduct GmbH, Germany.*

Klaus and Sigi's combination of Change Leadership and Kanban offers valuable insights and techniques for helping organizations learn and respond to change faster. I have found great success using these approaches over the last year and highly recommend this book for those in the field. *Cliff Hazell, Lean/Agile Coach, Spotify, Sweden.*

Kanban Change Leadership gives every change agent a set of tools to use when introducing a major change in an organization. Not only does it present the tools but also explains the underlying mechanisms that make the tools work. *Pawel Brodzinski, CEO, Lunar Logic, Poland.*

Want to meet the Peter Drucker productivity challenge? Then this book is a good start. *Håkan Forss—Lean/Agile Coach Avega Group, Stockholm, Sweden.*

"I have long respected Klaus and Sigi for their expertise in both Kanban and change leadership, and this book does a great job of integrating the two." *Mike Burrows, UK Director and Principal Consultant, David J Anderson and Associates, England.*

"This practical book explains Change Management with Kanban in a very deep and insightful way." *Vikram Sharma, Kanban Change Agent, Robert Bosch Engineering and Business Solutions Limited, Bangalore, India.*

FOREWORD

As I sit here at home in the summer of 2014 contemplating the next few years of for my business, my family, and my personal life, I realize that perhaps my children aged 12 and 9 will never actually learn to drive a car, unless they choose to do so for purely recreational purposes. The driverless car is now technically viable and commercial models may appear within the decade. The driverless car is truly a discontinuous or disruptive innovation. It will change the way we live and the next generation of adults will know a truly different lifestyle and society to the one I have grown up in. In a world of driverless cars, what is the difference between a family car, a rental car, and a taxi? Perhaps there is none? How will that disrupt existing businesses and existing lifestyles? What shifts in society will it enable?

Unlike the electric car, which uses an electric motor and lithium ion batteries, the driverless car is a truly disruptive and discontinuous innovation. The change to electric motors is on the other hand merely a continuous innovation. It doesn't disrupt the market; it merely offers an alternative technology for an existing system and operating model. It does shift the market for energy supply and it shifts the demand on natural resources, encouraging lithium mining rather than oil drilling, but life for everyday citizens doesn't change that much.

The technology industry tends to discard the value of continuous innovation—doing things better rather than differently or in a new and previously unavailable or unconsidered fashion. However, consider the world through the eyes of a 12-year-old. Children today assume that screens are touch sensitive and that they can interact with a device through the screen. They do not remember a world before the iPhone or the iPad. They've grown up in a broadband connected world. A rotary dial telephone is alien to them as is the concept that telephones are connected to the wall through wires. Continuous innovations can be easy to adopt because they don't require a huge change in lifestyle or business model. People with mobile phones simply upgraded to the latest model of smartphones and started using the touch screens, but the combination of ubiquitous broadband wireless data and high-definition color touch screens, available as smart handheld devices, did provide tremendous opportunity for change.

Continuous innovations—making things better, easier, and cheaper—commoditization, and democratization are all around us every day. Discontinuous, disruptive innovations come along every few years but the pace of these disruptive changes is accelerating. Each generation of technology innovation seems to help accelerate the pace of the next generation of discontinuous, disruptive change. Meanwhile, the world population is growing and the level of education is improving. More brainpower the world over means more knowledge workers producing more new knowledge and faster and faster innovations.

For the leaders of today's businesses this pace of innovation presents a huge challenge: business models that were solidly profitable for decades or even centuries and being threatened and disrupted. Uber is disrupting the market for taxis, a business model that is 120 years old, but only 5 years from now, driverless cars will disrupt Uber. Large businesses survive for shorter and shorter periods of time. Who would have thought, 10 years ago, that Nokia would have lost the mobile phone market and all but ceased to exist? Resilience is the new challenge for the senior leadership of big businesses.

How do you create a resilient business? How do you survive and thrive in a world that is moving so quickly where continuous innovations threaten your products and services on a monthly basis and discontinuous innovations threaten your business model every 2–5 years? The modern business must be able to change frequently and rapidly. It must have a core capability to enable and manage change. A modern business must be capable of evolving to adapt to a changing external environment full of new innovations.

When threatened with declining markets and possible extinction, businesses need to be able to experiment with new products, new services, and new models of service delivery. In turn, they may want to set the pace and stay ahead by producing their own, mostly continuous, innovations in products, services, and service delivery models. Innovations can provide a competitive edge. They can make the difference between surviving and thriving or declining and failing completely.

When a business wants to innovate, it almost certainly requires IT. IT projects are born out of a desire for a business to innovate or quickly catch up with a competitive innovation. IT projects are about change, and the service delivery from IT can provide a competitive edge if innovations can be delivered rapidly, at low cost, and with predictable outcomes. Equally, businesses can't always predict which ideas will work in the market and which won't. Evolving to stay fit for purpose requires experimentation—like generating several mutations of a species and waiting to see which produces the better result. As a result, there is almost infinite demand for IT projects because they represent experiments and guesses. The more bets we can place and the faster we can place them, the more chance we have of surviving and thriving. A strong capability in IT has become a core part of a strong adaptive, evolutionary capability in a business. Strong, fast, reliable, predictable IT services are now core to enabling resilience. More than ever, senior executives are pressurizing IT departments for more work and more services, delivered faster and with greater predictability and ideally at lower cost. When you don't know which options will work, you'd prefer to have lots of them at a low cost per option.

So, the new world of twenty-first-century business involves an ever faster pace of innovation and a greater need than ever to respond to a changing environment with new products, services, and service delivery models. Business must be capable of adapting quickly in order to remain fit for purpose. Resilience requires a strong ability to change. Change management is now a core business skill for survival in the twenty-first century!

The Kanban method was born out of the synthesis of two ideas to solve two different but related business problems. The first problem was tendency to manage IT projects in large batches and to commit too early to specifications when the requirements were still uncertain. In 2004, I introduced the concept of a virtual kanban system, with an IT department at Microsoft. The concept of kanban systems was adapted from Toyota's use of them in manufacturing industry. Kanban systems force deferred commitment, limiting the work in progress, preventing businesses from committing too early to things that are uncertain. Kanban systems force a discussion about what should be started now and whether we have enough information to start now versus what should wait until later and until more information is gathered or what should be discarded altogether. Kanban systems have a vital role to play in a world full of increasing pace of change and lots of uncertainty.

The second problem was a very human problem. People in IT departments were resistant to adopting new methods and processes. About a decade ago, I concluded that this resistance wasn't simply explained away as laziness or bad behavior; it was actually core to the human condition. The people were resisting adopting new methods and processes because they were wired to do so. So I asked myself, what would be easier to change, the design of the humans or the way we manage change and the introduction of new working practices? I concluded that we needed a new approach to change management: we needed an evolutionary approach to change.

In the English-speaking world, change management is dominated by the model developed by the McKinsey consulting firm. This model of prescribing a defined process or designing a new process to replace an old one, and then managing a transition from the old method to the new, has been around for over 80 years. It seemed to serve manufacturing industry and related physical goods industries such as distribution or retail rather well. Despite the ubiquitous nature of this model among large consulting firms working in the IT sector, I concluded that it was an obsolete model that wasn't compatible with the human condition. Trying to impose change on knowledge workers and creative people was simply a recipe for invoking passive–aggressive resistance. Instead, we needed a way to "start with what you do now" and evolve from there, and it had to be a way that engaged the people doing the work and made change and improvement an everyday concern for them. It had to self-motivate change from within, not some change imposed by change agents from the outside.

The industrial engineering model of the twentieth century was out, and self-motivated, management driven improvement was in!

In 2007, these two ideas came together in a synthesis that we now call the Kanban method. Kanban systems and visual boards are key enablers of a culture of continuous improvement. It turns out that visualizing invisible work on what has become known as a kanban board and deferring commitment through use of a kanban system are catalysts of employee and manager-driven process improvement. Change driven from the shop floor and undertaken as part and parcel of the everyday work of the organization produces just the experimental, evolutionary change mechanism we need in order develop a core adaptive capability for business resilience. Meanwhile, the use of kanban systems improves delivery through shorter lead times and greater predictability. Kanban Change Leadership becomes a core strategy for resilience in twenty-first-century businesses.

I'm delighted that Klaus Leopold and Sigi Kaltenecker's book has been translated into English. Klaus and Sigi hail from Austria, and their work in change management has been influenced by a number of German-speaking theorists, such as Baecker, whose work is not widely known or available in the English language. Klaus and Sigi open our minds to a different slant on change management by synthesizing ideas from the German-speaking world. Change management texts in English are all too dominated by the twentieth-century model popularized by McKinsey. So it is fitting that a book about Kanban, a new approach to change in the twenty-first century, should feature the diversity of thought in the field from experts not well known to English-speaking readers. Part 2 of this book offers you insights into the human condition and a deeper understanding of why a new approach to change is needed in the twenty-first century. It will help you understand how people take change personally and how to adapt your approach rather than push against the fundamental psychology and sociology of each individual. It will help you understand that the need for change is driven by change in the environment and to comprehend your business as a system that responds to that environment. This book will explain to you how and why Kanban offers us a new approach to change in twenty-first-century businesses using knowledge workers to do creative work and why Kanban Change Leadership can help even more traditional twentieth-century physical goods businesses by providing them with IT departments more agile and adept to deliver improvement and new capabilities faster and more reliably than before. For those of you who take the time to read this book thoroughly and internalize its contents, I have no doubt that it will make you more effective coaches and leaders of change using Kanban. Your business will gain more value from a deeper more effective Kanban implementation, and both you and your business should be better equipped to survive and thrive in this rapidly evolving world of work in the twenty-first century.

Seattle, July 2014 David J. Anderson

PREFACE

We are pleased to provide the English edition of our book on *Kanban Change Leadership*. As the kind appraisals show, this edition builds on encouraging resonance both from readers and from clients working on their specific culture of continuous improvement.

From both groups, we have learned a lot since the first German edition of our book had been published in 2012. In this regard, the current version is also a product of continuous improvement. For both the second edition in German and the English edition, we intensively reviewed the text, adapted new experiences, dropped some ideas, and changed others. The third part of our book, focusing on the practical implementation of Kanban, has probably undergone the most radical changes. Thriving on various lessons learned with many clients, we can now provide a simple four-phase model with clear goals and tried and tested tools.

As always, coming up with a new version does not necessarily mean to achieve perfection. The process of reviewing and rewriting our initial insights amplified a few question marks too. There are some limitations we were not able to overcome such as the emphasis of Kanban on team level, the rather weak focus on whole value streams, or the missing examples of portfolio or change management Kanban.

To effectively overcome these limitations, we would have to write another book—a book with a fresh approach to the broad field of Kanban, change and leadership, with different case studies, cocreating stories together with line managers and other key players, going even beyond the exclusive IT space, exploring the benefits of evolutionary change management in other business areas such as HR, finance, or graphic design. Unfortunately, we didn't find the time yet to realize this ambitious initiative—except for some current posts and articles that indicate what we have in mind for the future [1–5].

However, for the meantime, we wish you an inspiring read of this book—and a pleasant journey implementing some of the outlined change concepts and leadership tools. As always, we are pleased to receive all kinds of feedback.

Vienna, July 2014 KLAUS LEOPOLD AND SIGI KALTENECKER

PART 1

KANBAN

1

INTRODUCTION

"What should I do?" the Zen apprentice asks his master while standing in front of a tall ladder.

"You can climb the ladder, rung by rung, to the top."

"How many rungs does the ladder have?" asks the apprentice.

"Eighteen," the Zen master replies.

"And what should I do when I'm at the top?" the pupil wants to know as he places his foot on the first rung.

"You can stand there," the master explains in a friendly manner, "you can enjoy the view, you can climb back down, or you can continue to climb without any rungs."

This book has been written to give you the courage to climb further. It tells of ladders tall and short, of passionate climbers and spectacular climbs. A common feature of all climbs is that they begin with the first rung and then proceed step-by-step. Each one of these steps represents a small alteration through which you can gain new experience and improve.

We believe that this Zen story is a fitting introduction to a book about Kanban—after all Kanban is also about step-by-step change. Clear structures provide a gradual process of improvement that is relatively easy to establish. Many Kanban practices

Kanban Change Leadership: Creating a Culture of Continuous Improvement,
First Edition. Klaus Leopold and Siegfried Kaltenecker.

are like simple ladders. It is due to this that Kanban is quickly becoming a sensation, enjoying widespread popularity in the world of software development.

"Kanban rocks" is how one of our customers summed it up. He, like many other Kanban fans, has reason to be thrilled. Kanban:

- Follows **simple rules**
- Is built and runs on **easy-to-master mechanics**
- Can be implemented with **relatively little effort**
- Can lead to **remarkable improvement in very little time**

Sounds good, doesn't it? However, we have not written this book just for the growing Kanban fan base. We will emphasize critical aspects and the several traps into which users repeatedly fall and present some practical guidelines for Kanban change management to help avoid these traps. In order to do so, we will investigate various starting points, identify relevant system and environmental factors, and describe the personal challenges involved in a process of continuous improvement. Ultimately, Kanban is always about the whole system. Kanban:

- Often starts with a small team but **always has its eye on the organization as a whole**
- Concentrates on technical development but is simultaneously **always aligned with economic value creation**
- Aims to improve software development processes but **requires everyone involved in these processes to be willing to change**
- Is quick to apply but **requires mindfulness in order to improve continuously**

It is relatively easy to start a Kanban initiative at your place of work. However, it is highly challenging to implement the initiative in such a way that you create a culture of continual improvement. Practice shows that a *quick-fix* approach to Kanban at the workplace will rarely deliver long-lasting change—professional change management is required to achieve a sustainable environment.

1.1 WHAT WE CARE ABOUT

Kanban Change Leadership will show you all that is necessary to properly understand change management with Kanban and be able to apply it optimally. In order to achieve this, we provide you with many maps, tools, and, most importantly, various scenarios. We draw on our own experience as Kanban coaches and change experts to enable you to read real case studies and then apply what you have learned systemically. In other words, we attempt to smuggle valuable knowledge about organizations, cultures, strategies, and emotions from systems theory into the book without losing sight of the real world. What good is the best theory in the world if you're not capable of applying it appropriately?

On the subject of appropriate action, in a study carried out by Kimberley-Clark, people were asked what they would take with them on a desert island. More than 50% of the 1000 people asked said it would be very important for them to take toilet paper. What can we conclude from this? As the German economist Günther Ortmann put it, "people think practically" [1].

Practical thinking is a requisite for twenty-first-century change management. In this book, thinking is based on four fundamental principles as stipulated by David J. Anderson [2]:

1. **Kanban begins where a system is already in place**. No big change, rigorous training or process transformation is required. You have already begun climbing the Zen ladder simply by bringing about awareness of your current work processes.
2. **Kanban respects the current state**. Neither the current processes nor the existing functions are called into question. In this context, to respect is to assign meaning to that which is already there and subsequently, together with all other value-creation partners, build on this meaning.
3. **Kanban seeks incremental, evolutionary changes**. It's all about proceeding step-by-step—not in a single, massive leap—and agreement among all essentially involved in this process of change. In other words, Kanban requires that all stakeholders in a given value-creation process have a shared understanding of the work and improvement, regardless of whether this concerns the core team, clients, suppliers, owners, or senior management.
4. **Kanban requires leadership at all levels of the organization**. In order to create a culture of continuous improvement, all involved should contribute their ideas for improvement and be able to implement them. The operationally active employees frequently best know what needs to be improved in their daily work environment—let us support them in equalizing their viewpoint with that of management and taking the next step toward improvement together.

We believe that beyond these principles a profound fundamental understanding of how a culture can create continuous improvement is necessary. Our opinion is that the following principles are relevant:

1. **Kanban is an initiative for change**. We are concerned with systemic improvement, where collaboration rather than individual performance is important. Value creation and quality of work increase due to better structures and clearer rules of play between all cooperating partners.
2. **Kanban is concerned with the overall working environment**. The improvement of this environment requires critical reflection on each individual's fundamental mindset, expressed in terms of performance and cooperation. This in turn requires the willingness to continually work on one's self-development.
3. **Kanban revolves around people and not around mechanisms**. It is people who drive a sustainable process of improvement, and they achieve this very

visibly through emotions: joy, courage, enthusiasm, but also anger, disappointment, and sadness. We strongly recommend that these emotions be respected and used since, ultimately, they can very much be seen as the key drivers of change.

4. **Kanban is a team sport**. You need allies to create a culture of continual improvement. You need partners who will create and sustain new value with you. You need the support of your management because you want to expose systemic problems and resolve them. And you must have your stakeholders on board because you cannot create the added value you want without their active cooperation.

These principles emphasize the complexity of the change you can effect with Kanban. It requires an approach to match this complexity, and this is the reason why simply diving into Kanban is not generally recommended—you would risk achieving short-term change at the cost of the long-term potential for improvement. In the context of the introductory story, you would climb down again after reaching the tenth rung if at all and never get to the point where you climb further, without rungs.

"It shows who is truly committed," a colleague once said in a discussion about this limitless climbing. Be sure of your decision before making such a commitment. Use our guidelines to define your point of departure before embarking on your Kanban adventure. Try to identify the corporate culture you belong to. And assemble a training program tailored to your personal work situation from the exercises we provide.

1.2 WHO SHOULD READ THIS BOOK

There are three target groups we particularly want to reach with *Kanban Change Leadership*:

1. **Those who are fundamentally interested in Kanban**: "Hey, this is cool! What is it exactly? How does Kaizen work?"
2. **Those involved in change management in IT**: "What approaches are there? What are the unique features of a process of continual improvement? What can I personally adopt from Kanban change management?"
3. **Those considering a Kanban initiative or already underway**: "What do I have to look out for? How do others do it? What could I also try out?"

The three parts of this book correspond to these three target groups.

In the first part, we focus on the foundations of Kanban. What are the basic assumptions? How can you visualize the current situation? What is the purpose of work-in-progress (WIP) limits? What are service classes? How can you apply metrics? And much more. Part 1 establishes the technical basis of Kanban and indicates the mechanisms required.

In the second part, we explain the context of Kanban change management. What are the options for change? What can they set in motion? What are the consequences for a business? What particular opportunities does Kanban-driven change provide? Besides mechanistic formulas and processes for automatic improvement, in Part 2, we share with you a contemporary understanding of the professional process of change. Despite the fact that everyone talks about change, there is still plenty left to say on the matter.

In the third part, we relate the technical system of Kanban with the social system of business and show you, using selected case studies, how to build a culture of continual improvement. How do you start the process? How do you define your point of departure? How do you create a Kanban system tailored to your field of work? What should you look out for when using it? Part 3 provides you with a compendium of experience showing how Kanban is applied in various situations.

"I don't know whether it will be better when it's different," the German philosopher Lichtenberg once said. "But that it must be different to be better, that much I know." In this spirit, we wish you a most inspiring read and good luck.

2

KANBAN PRINCIPLES AND CORE PRACTICES

Do you know the film *Modern Times* by Charlie Chaplin? There's a famous scene where the chairman of the corporation suddenly orders the speed of the conveyor belts to be increased without warning, "Conveyor belt five is running too slowly. Double the speed!" Chaplin struggles as best as he can, trying to tighten all the screws as the products speed past. However, he always ends up falling behind—at this speed, it only takes a short sneeze for him to repeatedly get out of sync. Consequently, Chaplin is forever getting in the way of the next conveyor belt worker, who is hammering, disrupting the entire process. Colleagues and supervisors drag him back to his allotted position, but it's no use—he simply can't keep pace. And then it happens: in a wild frenzy of tightening screws, nobody can stop him any longer, and he is pulled onto the conveyor belt and swallowed up by the machine driving it. He elegantly glides between the gigantic cogs and is spat out by the machine on the return run. His accident has left its mark: he suddenly wants to tighten anything remotely resembling a screw, including his colleague's nipples and the buttons on a secretary's skirt.

Released in 1936, Chaplin's film was a harsh critique of the prevailing assembly line conditions. The following appears in the opening credits:

> Modern Times: a story of industry, of individual enterprise—humanity crusading in the pursuit of happiness.

Kanban Change Leadership: Creating a Culture of Continuous Improvement, First Edition. Klaus Leopold and Siegfried Kaltenecker.

How would Chaplin stage this film today? For most of us, today's conveyor belts are desks and computers; we have better salaries now and the workers in the 1930s could only dream of the social benefits we enjoy. But in the present day—shaped as it is by knowledge work—there's always someone who still shouts, "Double the speed!" It seems that history is repeating itself. Today, knowledge workers also have to strive for good working conditions. It is no longer about the basic demands for a humane working environment such as light, breaks, and safety, but rather about the issue of time and the right to not be required to be available for the company around the clock. In a nutshell, it's about the ability to complete the work at hand within the allotted time period. Of course, there is also the flip side of the coin: global competition is often the source of huge amounts of pressure, causing organizations to reduce the two factors—quality and speed—to a single issue. Is it even possible to bring together this issue with the demands of a socially acceptable working environment? Can one be more relaxed at work while being more productive? We say yes, it is doable when you create adaptive systems in which people are able to find their own way to improvement.

2.1 SEEKING PRODUCTIVITY

Industrial assembly is all about the economic principle of optimizing the process between the amount applied (input) and the amount yielded (output). "Act in such a way that the desired accomplishments are achieved with the minimum means (the minimum principle), or, rearranged, that the accomplishments with a given amount of means are as high as possible (the maximum principle)," states Zäpfel [1]. One is thus seeking the highest productivity possible.

The idea of optimization is often so incorrectly interpreted that it suddenly states, "Use as little as possible to achieve as much as possible." Ironically, it is precisely with this perspective that we are commonly confronted in the practice of knowledge work. With unaltered processes, structures, and resources, as much input—that is, tasks or jobs—as possible is crammed into the system in the hope that as much valuable output as possible will emerge at the end.

Peter F. Drucker, one of the pioneers of modern management education, anticipated this problem 20 years ago. In his 1991 article "The New Productivity Challenge," he demonstrated how productivity in "making and moving things" has constantly increased since the onset of the industrial revolution and how this has developed and continued to nurture the well-being of (above all) Western society. Nowadays, Drucker said, productivity continues to increase steadily, but the great revolutions in production, mining, construction, and transport have already happened. He explained that the workforce has shifted from the classical areas of production into the sectors of knowledge work and services. Drucker therefore asserted right at the beginning of his article [2]:

> The single greatest challenge facing managers in the developed countries of the world is to raise the productivity of knowledge and service workers. This challenge, which

> will dominate the management agenda for the next several decades, will ultimately determine the competitive performance of companies. Even more important, it will determine the very fabric of society and the quality of life in every industrialized nation.

Today, we know just how right Drucker was. Knowledge-intensive sectors labor endlessly to find the button to press or the screws to tighten in order to increase the productivity of their knowledge workers. Interestingly, there are very obvious parallels between questions of optimization in industrial production and knowledge work—this will be seen in Kanban's individual steps. Equally, however, there are very sharp contrasts between the two sectors.

But how does one define knowledge work?

KNOWLEDGE WORK

German systems theorist Helmut Willke [3] describes knowledge work as follows:

> Nearly all human activities are based on knowledge in the sense that experience and knowledge play a part. Practically all forms of skilled work, above all the classical professions (doctor, lawyer, teacher, academic), are knowledge-based forms of work, based on specialized expertise that these professionals must acquire through extensive processes of education and training.
>
> The concept *knowledge work* means something else. It describes practices (communication, transaction, interaction) in which the required knowledge is at no point in life acquired simply through experience, initiation, teaching, skill training or professionalization—nor is it implemented in this way. It is much more the case that knowledge work in this sense requires that the relevant knowledge
>
> - Be continually revised;
> - Be permanently viewed as capable of improvement;
> - Be observed not principally as truth but rather as a resource, and
> - Be indivisibly associated with ignorance, meaning that certain risks are unavoidably connected with knowledge work [3].

Manual work thus differs from knowledge work since ignorance—and its necessary reflection in knowledge work—constitutes a dimension that is hard to influence. The underlying problems, that is, the exercises and tasks, are also significantly more multifaceted in sectors such as software development than they are in the assembly of—in the literal sense of the word—*tangible* products. In these sectors, it's much more frequently about inventing something completely new or refining something that already exists rather than simply reproducing something already established. Simply put, *thinking* and *problem solving* can't be easily standardized.

However, just as there is a big difference, there is also great similarity. Regardless of whether one is developing software or constructing a car, the person carrying out the task should in both cases have the ability to complete certain stages before beginning new ones, irrespective of the process as a whole.

We only need to look at how we carry out practical work in our daily life to see this. When we build shelves, for example, it is clear to us that we should perform the various actions sequentially. Only a very few of us are able to use a hammer and a screwdriver simultaneously. We complete the steps one after the other and concentrate on a task at a time.

Strangely, this logical perspective with regard to the completion of tasks disappears when it comes to knowledge work, where it is often assumed that many tasks can be simultaneously carried out by the same people. More so, tasks that have nothing to do with the actual core goal (e.g., excessive administration) drift into the "production area" of knowledge work. In contrast to production companies, merely pumping more money or technology into a process doesn't result in a significant increase in productivity either. In knowledge work, the only possibility is to work "smarter"—as was also Peter F. Drucker's understanding—meaning that one focuses only on that which is absolutely essential [2]. For Drucker, the foundations for a smart work ethic lay in the answer to the questions: What is the task? What are we trying to achieve? Why do we need to do it at all? The greatest increase in productivity in knowledge work is achieved when we define the tasks and goals clearly and only do the things that are absolutely necessary.

PRACTICAL KANBAN

The seminar in Zürich with David J. Anderson had just begun. Everyone was there to learn how kanban works, and of course, we had loads of case studies up our sleeve—for practical experience. All of a sudden the fire alarm rang. It wasn't a drill; somewhere in the building, there was a real fire. So the whole group followed the escape route out and into the neighboring café. After a brief moment of panic due to the onslaught of a hoard of homeless seminar attendees, the baristas did the one logical thing. "Coffee?," one of the employees called out into the crowd, and a small group of coffee addicts formed from the disorganized rabble. Those wanting food were organized into a second queue, and whoever didn't want anything just sat down. Logical, simple, efficient, and pretty smart. Why did the baristas react so quickly and efficiently? Well, because they simply knew the bottleneck in their process—quite obviously the coffee machine. Using this knowledge, they were able to adapt their modus operandi as quick as a flash. Were it not the fire, we couldn't have hoped for a better introduction to the seminar.

2.2 kanban and Kanban

A further curiosity of the application of knowledge work is that an individual is often seen by many as a factor that needs to be optimized. Organizations therefore initiate expensive further education programs and invest heavily in keeping the knowledge level of their employees as up to date as possible. Fundamentally, this is to be praised, but it disregards one thing: even when an employee knows everything there is to know in his/her field, that doesn't necessarily make him/her or his/her team any quicker. Despite all, one can only accomplish a certain amount of work within a given period of time. If you only ever want to optimize an individual, you're failing to take into account something William Edwards Deming put very succinctly [4]: 94% of the performance of an organization is dependent upon the conditions of the system, and only 6% is dependent upon the employees. According to Deming, every significant improvement in quality and productivity is a result of measures that deal with the system. Just like Drucker, Deming says that employees should be helped to work smarter rather than harder.

The most renowned example of permanent change and improvement to a system is the Toyota Production System (TPS). The reason that Taiichi Ohno and Kiichiro Toyoda worked so intensely on the improvement of their production system is that theirs was a similar situation to the one seen today in knowledge work, defined as it is by one-off production. In the case of Ohno and Toyoda, the market demanded many different car models in small quantities. Diversity on this scale was no longer achievable with the production model that Henry Ford had perfected. Ford had achieved cost efficiency with his radical division of labor, however without the possibility to adapt the best-selling "Tin Lizzie"—the Ford Model T that was revolutionary for its time—for special requests. Ohno and Toyoda realized that the problem of variety production could not be solved by simply burdening employees with an even stricter, more monotonous division of labor. They further wanted to deliver the best quality at low cost and within the shortest possible processing time.

So they looked at the issue from a new angle, concentrating on the movement of the product through the entire production process. They satisfied the constant demand for an increase in productivity with the principle that only that which is really necessary should be done, precisely at the point in time it is needed, and in the required quantity (just in time (JIT)). This also concerns the avoidance of waste. Toyota in fact defined three different types of waste [5]:

1. Tasks that use up resources without supplying any additional value (*Muda*)
2. Irregularities (or too high variability) in the production process (*Mura*)
3. Overload (*Muri*)

The goal of built-in quality is achieved via *Jidoka*, the instant identification of errors and problems. Production is stopped instantly the moment an error occurs, because experience shows that errors that aren't corrected go on to appear in other areas as well. The core elements of the TPS production-process control are **kanban**—*kan* is Japanese for "visual" and *ban* means "card." These *visual cards* in downstream production stages indicate that a task has been completed and a replenishment of assembly components or material is required in order to be able to continue working. This pull system reduces inventories to a minimum. Simultaneously, problems in the production process become immediately apparent when the assembled products suddenly pile up in the upstream stages of production. The trick behind this is to limit the number of **kanban**. You can only feed as much work into the system as the available visual card permit.

In his search for improvement possibilities for software development, David J. Anderson, the pioneer of Kanban for IT, indirectly came across the TPS. During his first deliberations, he started off primarily with the concept of the "drum buffer ropes" in Eliyahu M. Goldratt's *theory of constraints*, which, put simply, ascertains that every system has specific bottlenecks that limit the possibilities for value creation. This is because the bottleneck determines the rate of flow (this will be discussed at greater length in Chapters 4 and 7). The thinkers at Toyota had already realized this decades ago and believed the simplest way to optimize the flow was to let the bottleneck itself determine how much it could currently process. Kanban in IT brings together the best from the most varied intellectual approaches. However, this will be initially paired with and developed along practical experience such as the approaches of evolutionary change, making rules explicit, or the classes of service. We will look at these more closely in subsequent chapters. Kanban in IT is therefore not the transfer of an individual concept from industrial assembly to knowledge work but rather a hybrid of concepts. It is easy to explain why the term kanban has become so established: it reflects the most important core points, is intuitive, and is easily pronounced by people the world over.

TERMS USED IN THIS BOOK

kanban: A kanban is literally a tag that not only enables but also ensures JIT production. Seen as a totality, it is a system of time management for production companies that helps decide what, when, and how much is to be produced. In knowledge work, we use a virtual kanban system as a means to represent work items.

Kanban: The evolutionary change management method developed by David J. Anderson. It supports change in an evolutionary sense by successively optimizing existing processes. We use a capital "K" when referring to the Kanban method in order to distinguish it from the production kanban and the virtual kanban system.

What do we mean by system?

System in ancient Greek means "body, organized whole, that which is connected." In contemporary sociology, it describes a meaningful unity of elements that differentiates itself from the surrounding environment. According to Niklas Luhmann, who is considered the father of sociological systems theory, "a system is an organised complexity" [6].

Social systems are complex bodies produced and reproduced via communication. Society and all its organizations and interactions are "communication network(s)" [7]. This makes them living beings but also incalculable.

Psychological systems operate in the form of processes of self-awareness that can be described as a meaningful unity of perception, thought, feeling, and desire. They are inseparably connected with social systems although they are not a part of them.

Technical systems unite elements whose interaction likewise forms a unity. This interaction-based unity however is not defined in terms of meaning, but rather in terms of function. It is highly structured and mathematically predictable just like a computer or operational system.

Kanban systems (capital "K") describe the complex interrelation between social, psychological, and technical elements geared for continual improvement. Kaizen—the Japanese term for "change for the better"—demands a goal-oriented bond between the organization, the employees, and the work processes.

By **technical kanban system**, in the narrow sense, we mean the form of visualization of the work process (e.g., via a board) and the individual instruments (e.g., tickets, meetings) that help provide insights into your own processes. The visualization simultaneously indicates the specific individual in the particular value-creation chain we wish to optimize. The most important characteristic of a technical kanban system is that it quantitatively limits the work in progress.

Kanban team, team, or Kanban group refers to all those who work with a Kanban system and actively apply Kanban practices. A group of this nature does not have a fixed size but rather changes as the application of Kanban progresses. It can increase or decrease in size and can consist of people from the most varied areas, departments, or teams of an organization.

What do we mean by "stakeholder?"

In German parlance, the term "stakeholder" is mostly used to denote an "interest group." In a corporate context, it refers to all who have a certain input or concern in an organization. This extends the corporate management definition of the purely economical understanding of stakeholders (an owner or holder of shares) to the extent that it includes social, cultural, and ecological interests.

In a systemic sense, one distinguishes between internal and external stakeholders. Strictly speaking, when considering factions, the employees, managers, and owners belong outside the corporate system, just like customers, suppliers, business partners, creditors, countries, or NGOs.

Therefore, **in this book**, stakeholders of a value-creation process are always those either directly participating in the process or perceptibly affected by its consequences. The identification, addressing, and mobilization of the stakeholders together constitute one of the central foundations of the Kanban change leadership model we have developed.

When we speak of **customers** in this book, we refer exclusively to internal customers or the internal representatives of external customers.

2.3 EVOLUTIONARY CHANGE MANAGEMENT

Today, organizations both in the field of software development and beyond try to be agile and lean in order to succeed in the global market.

But can one simply become agile and lean from one day to the next?

Let us once again return to the TPS. The critical cultural component of the TPS is the notion of continuous improvement, *kaizen*. All measures taken in the production process are themselves expressions of this notion or indeed of this inner attitude. In Ford's assembly plants, workers simply had to become ever quicker by concentrating exclusively on one action, possibly ending up as deranged as Chaplin in *Modern Times*. If it is only ever the speed of an action that is given emphasis, there would be no remaining time for thought on improvement—you're trapped in monotony. Toyota's kaizen culture gave employees back their time, responsibility, and the possibility to identify problems and suggest possible improvements. This brings us to an important observation that change comes from the people themselves.

Change comes from the people themselves.

We don't consider Kanban an agile method for software development in the conventional sense. It's not about applying methods, be they Kanban, Scrum, or XP. We believe that the primary task of an organization is to be successful. Kanban is one possible means of helping make organizations successful. But methods alone will not suffice—workers need to imbue them with life. They must understand which problems can be identified using Kanban and proceed accordingly.

In this sense, Kanban's path comprises small, continuous, evolutionary steps. Kanban is no magic spell that will conveniently create success. Success is the result of changing the way you think and act. The essential difference between Kanban and many other agile methods is the **evolutionary nature of the change**. Systems and processes are created for particular reasons, or they exist in their current state due to development that took place for particular reasons. The development of an organization is always intimately connected with external factors that cannot be influenced directly—due to dependency on legally prescribed standards such as those in the aviation industry, for example. However, despite these restrictions, every organization can and must find a way to remain competitive by continuously improving its own processes.

In classical change management, process engineers would go about looking for the optimal new system for a corporation. They would initially develop a "big picture": an image of how all sequences of events should be shaped in order to be considered optimal and satisfy the most varied requirements. This is normally a lengthy process that stipulates a target level based on the current state of knowledge for a given near or distant point in the future. A second weakness is that these target-level deliberations occur mostly in isolation, far removed from the people who will be affected by the change and know the problems of their daily work environment as well as the weaknesses in the system perfectly well. In the best-case scenario, affected employees can voice their opinions to an extent but must then live with the resulting decisions and forcibly "improve." At one point, the change philosophy of the organization will be officially imposed in the hope that everything will—magically—start to improve. In such a situation, organizational change would only be about the result rather than what might be wrong, disregarding the fact that the change is only ever valid for the present and not for eternity. Something else that is also disregarded is that change always induces fear.

It is not the case that every change is an improvement. The larger the change, the greater the fear is. And this fear cannot be assuaged by logical argumentation about the purpose and utility of the change. The simultaneously growing danger is that initial successes will remain a flash in the pan, eventually disappearing due to active or passive protest by employees who have been taken by surprise. Kanban's goal therefore is not that of establishing a loosely defined, "optimal" mode of operation. *It is far more concerned with remaining constantly on the lookout for things that can be improved. The goal is rather to develop a kaizen culture step-by-step, a culture that is focused on providing better results for the organization in economic and, for those people working for the corporation, social terms* [8].

Kanban operates under the assumption that the power to make improvements and develop further is already inherent in the existing processes. It begins with the current processes and cycles. No target levels for the distant future are defined, since not everything about the existing process is bad. The conceptual core is installing into the system the mechanisms that make continuous change and improvement possible. In contrast to classical change management, the path in this case evolves *through* walking it. All those participating should themselves identify where problems are to be found, what they can do, how they can help themselves, and when and how they must act in order to prevent misunderstandings. Just because the current situation is the starting point doesn't mean that Kanban is an excuse to maintain the status quo. Change must always take place.

Even though Kanban and its tools initially look pretty simple, the difficulty lies in anchoring the notion of kaizen in the value system of the people involved.

KANBAN

Kanban is a complex, adaptive system. The fundamental principles according to Anderson are:

1. Start with what you are currently doing.
2. Pursue incremental, evolutionary changes.
3. Respect initial processes, roles, responsibilities, and job titles.
4. Encourage leadership at all levels of the organization.

Neither in its entirety nor in its constituent elements is Kanban simply about getting from State A to State B with the greatest amount of effort (which often doesn't even work anyway). The system is optimized in a series of small steps in order to initially reach A′ securely. When A′ has been reached, you can proceed to A″, etc. (Fig. 2.1). And this is precisely what a Kanban group should be measuring: the first small step.

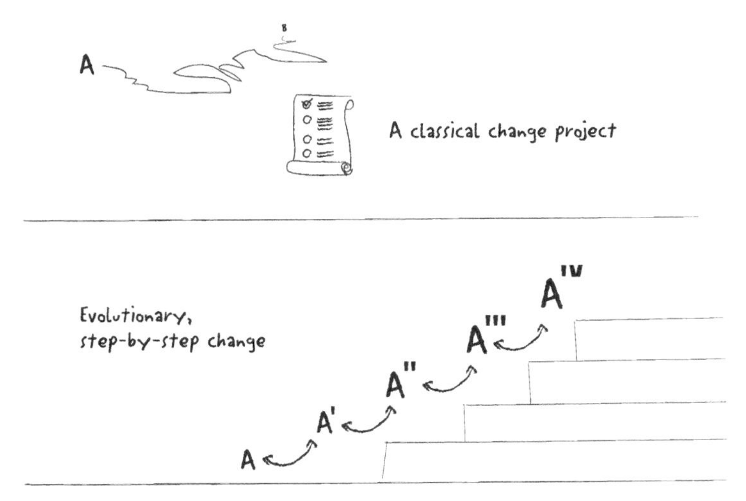

FIGURE 2.1 A classical project of change versus evolutionary change.

The constant question is whether an evolutionary change such as the one Kanban represents might not in fact take a great deal of time. The answer, probably unsatisfying to most, is: yes, it can, but the opposite could also apply. We have had experience with teams in finance that grew from a few Kanban beginners to a group of 40 people within 3 months, who coordinate their work with Kanban boards. These people speak actively and directly with each other—people who previously didn't even know each other by name. Most importantly, they discuss their work together in the daily standup meeting of their own accord, not when someone prompts them to do it. They have quite simply come to the conclusion all by themselves that these conversations are beneficial to their processes and results.

2.3.1 Knowledge Work: The Problem of Invisibility

At the beginning of the chapter, we discussed how classical production and knowledge work are similar to each other in some respects and differ from each other in others. One of the greatest weaknesses—if you like—of knowledge work is that one cannot see what actually takes place during the production process. The assembly line is in the heads of the employees, and as we know, one cannot standardize or control goal-oriented thought processes.

We are hence faced with two problems:

1. Since internal and external employers cannot see the intellectual production process, it could be that software developers are required to do several things at once.
2. Software developers themselves are equally unable to see what is going on in this production process. They walk blindly into a trap, saying that they're perfectly capable of completing several tasks at once, and mostly always

> manage somehow. However, the question is: at what personal price and at what cost to the organization? One can easily identify the bottlenecks in production assembly lines: precisely at the point where the partially completed products queue up. In knowledge work, this transparent overview of the work being done is not available. We identify bottlenecks by our own panicked reactions but lack the representation of the entire process necessary to understand the areas we need to work on so that we can effect change and improvement.

The process of knowledge work must be brought out of the darkness and displayed in order to find starting points for changes and improvements. It is only the illumination and identification of this work that creates awareness of the limitations and possibilities in terms of capacity.

2.4 KANBAN CORE PRACTICES

This brings us to the core practices of Kanban—the points that must absolutely be considered in order for this adaptive system to function properly. Fundamentally, Kanban stipulates very little in terms of *how* something should be done; it is more the case that Kanban makes suggestions *that* something should be done. Stipulations would be counterproductive since it is mostly the people working in the system that should recognize what needs to be changed and what form this change should take. According to David J. Anderson, there are six core practices that make a Kanban implementation successful:

1. Make work visible.
2. Limit work in progress (WiP).
3. Manage flow.
4. Make progress policies explicit.
5. Implement feedback mechanisms.
6. Improve collaboratively (using methods and models).

2.4.1 Making the Work Visible

The goal of Kanban is to establish a continuous workflow that ultimately generates more value for the customer. Kanban helps make visible the processes of knowledge work, thus also the associated problems that limit the workflow. Initiating quantitative restrictions (WiP limits) for the work makes clear what causes the system to falter and what hinders the completion of operations. This is precisely what happened to the team mentioned earlier: through the visualization of their workflow, it became clear that certain people in the organization should communicate directly with each other in order to improve their processes.

Moreover, the crucial difference between Kanban and other popular modes of operation is that work doesn't simply get "passed on" to the next stage of processing as soon as a team member is finished with it (the push principle). It is much more the case that team members of subsequent stages collect the work from the upstream stages as soon as they have the capacity to do so (the pull principle).

2.4.2 Limiting the WiP

Traditional production management teaches us that unfinished products are tied-up capital. This is why every production organization tries to keep the number of half-finished products as low as possible. Again, we have the problem that the value of tangible products is easier to quantify than that initially resident in the minds of software developers. In any case, the following is true for both traditional assembly and knowledge work: the greater the number of active operations in the system, the higher the lead times. Let us demonstrate this using the example of allocation. All that has been assembled and is ready to be delivered can be invoiced, earning the organization money. From an economic perspective, it is thus cleverer to carry out one operation 100% of the way rather than 10% of each of 10 operations. Therefore, in order to reduce the lead times and establish a continuous workflow, it is sensible to limit the number of operations carried out simultaneously at any given stage. Hence, we talk about limiting the "WiP" or simply about "WiP limits." The relationship between the number of simultaneous operations and the lead times becomes clear in Figure 2.2.

The graphic shows how three operations can be arranged sequentially or quasisimultaneously. "Quasisimultaneous" means that one constantly switches between the three tasks, since people are strictly speaking incapable of carrying out several active tasks simultaneously. In the sequential process (the three large blocks in the

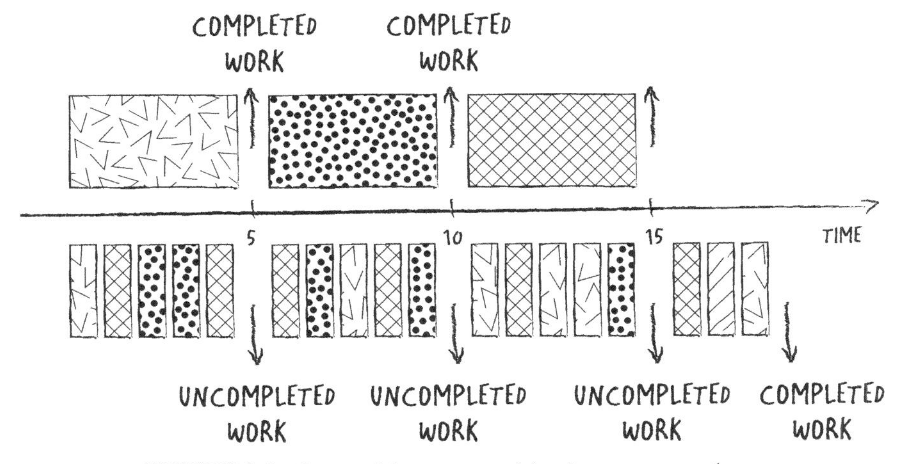

FIGURE 2.2 Sequential versus quasisimultaneous operations.

graphic), one can see that the tasks are processed in 5 units of time, respectively. In the quasisimultaneous process, the lead times increase to 16, 17, and 18 units of time. The continuous change between tasks requires additional effort since the project members must always refamiliarize themselves with the new task—in a classical production process, this is the changeover time for the readjustment of the machines. In this simplified example, the additional effort is quantified as 1 unit of time per task.

Together with minimizing tied-up capital in the process and reducing lead times, limiting the WiP brings a further advantage that is directly related to the goal of continuous improvement. In a WiP-limited pull system, the bottlenecks are made visible—the employee involved in the bottleneck cannot collect any work from his colleague one stage upstream because he is still busy with the current task, and since this colleague is limited by a WiP limit, he cannot collect any work from *his* colleague a further stage upstream. The consequence is that the workflow becomes blocked and employees become incapable of further work. WiP limits are therefore in principle a requirement to enable a pull system as envisaged by the principles of lean production to evolve.

Probably, one of the most important advantages of WiP limits concerns the relationship with customers. Not honoring commitments is a constant burden on the relationship with the customer—be it an external customer or a colleague in the organization—and can damage the image of the organization in the long term. In the world of software development as well as in other fields, it often occurs that operations run simultaneously and one simply runs out of time. One normally then tries to defer meetings or intentionally make reductions in quality. Often forgotten at this point is that any reduction in quality will reappear further down the line, in the form of complaints, faulty products, or requests for alterations that disturb the current work process yet again. With Kanban, the goal is to make only those commitments one can honor. It can therefore also be appropriate to say "no" when additional tasks would break the WiP limit.

2.4.3 Managing the Flow

The Kanban focus is on the workflow. This means that everything that hampers the flow of work, such as blockers and bottlenecks, receives particular attention. The motto is: work on your problems first before going on to new work. Furthermore, we want to be able to honor agreements we make with our partners enabling us to develop our mutual trust. In order to keep promises and honor agreements, we must know precisely what we're capable of achieving. When we change something, we also want to be able to see whether the changes prove to be of value. We therefore must measure the extent of our approach to the stated goals. However, this doesn't mean that we measure the performance of individual employees but rather the performance of our kanban system. We want to examine whether we have shaped and developed our capacities and processes in this segment of the value chain in such a way that they can deal with the required tasks. If they can't deal with the tasks, we must change them again—this is the basis of a kaizen culture.

CHANGES TO BEHAVIOR VERSUS CHANGES TO THE SYSTEM

We can approach problems with the workflow from two perspectives. Take, for example, quality problems in writing software. One possible reaction is: "Code better!" This only addresses the skills of the employees. What's not taken into consideration, however, is that employees' performance is also dependent upon their environment, that is, the influences and disruptions that buzz around the system they are involved in while they're working.

Another possible reaction would be: "Apply test-driven development. First, write a test case and then program afterward." Alternatively, you could shorten the feedback loop between programming and testing. In this case, we are concerned with a systemic change: we rebuild our social and technical systems in such a way that we can deliver better quality.

Kanban aims to establish a fast, predictable, and consistent workflow. To control this workflow, we must first identify which work item types a team must generally complete and be clear with ourselves that not all tasks have the same level of urgency. We therefore introduce classes of service for a Kanban system. This principle is apparent in parcel delivery services, for example, where various levels of service cater for the urgency of the delivery: the routes the drivers take are planned accordingly.

Work item types and classes of service are the basis for the service-level agreement (SLA). This is how the team delivers on the guarantee to accomplish operations of a particular class of service or a particular work item type within a defined time period. It provides the stakeholders with a high level of strategy security, because teams normally exhibit an SLA reliability of over 90%.

Communication is the common link between all measures related to control and the measurement of the workflow. The so-called daily stand-up meetings, where team members discuss their work in the presence of the board, are important for teams. The goal of these meetings is to coordinate operations and maintain the workflow.

2.4.4 Making Policies Explicit

The work style of a Kanban team can be seen as a collection of policies that the group imposes upon itself. These policies are to be transparent to all participants and must be adhered to and have two effects:

1. Errors and aspects of a policy that sometimes need to be changed can only be identified when both the team and the stakeholders adhere to the policies. The training course we relocated to the neighboring café is an excellent example. The baristas immediately realized that the policy "everyone must stand in a single line, regardless of the particular type of order they would like to make,"

was no longer feasible when 25 rather than five customers suddenly congregated at the counter. One of the first policies states that a problem should be solved as soon as it appears. The policies themselves are no exception to this: if a policy is no longer effective, it should be changed. The moment we stop changing our policies and standards, we halt the process of improvement.

2. Policies remove a large portion of emotion from discussions. One moves from subjective accusations of guilt to significantly more objective discussions about policies, making consensus on subjects in dispute more likely. This effect is not always immediately apparent. Initially, due to established habits, a guilty party is often sought when a policy isn't adhered to. At the very least, someone should assume the responsibility of raising conversations to the pertinent level, where the policies themselves and not the individual people are the subject of the discussions.

2.4.5 Implementing Feedback Mechanisms

With Kanban, everything is focused on achieving continuous improvement. Learning plays a decisive role. In order for us to be able to learn something, we need feedback to identify what we can do better. For example, many organizations opt for daily stand-up meetings where feedback about the current work situation is exchanged. We recommend that these meetings take place not only at the team level but also on a larger scale, encompassing the entire value chain. The number of participants can be restricted by using delegates of individual teams to coordinate those of other teams in front of the board. Retrospectives, targeted improvement meetings, are also an important feedback mechanism. The following maxim applies: the broader the spectrum of participants, the better the feedback. Operational reviews are often used for organization-wide learning about metrics. Whatever helps establish high-quality feedback about the actual process should be integrated into the everyday working practice in order to learn and improve.

2.4.6 Carrying Out Collaborative Improvements

Improvement doesn't mean that we constantly have to reinvent the wheel. In the case of multiple problems, we can fall back on approaches and models that illuminate the sets of problems that constantly reappear in all systems and have hence already proved their value in practice. Kanban is thus itself an adaptation of available practices and ideals—for example, from the automobile industry—for the purposes of software development in particular and knowledge work in general. David J. Anderson found several fitting and well-proven theories for the basic principles of Kanban such as Eliyahu M. Goldratt's aforementioned theory of constraints; economic understandings of waste in the form of transaction and coordination costs, for example; or the influences of variability on a given system.

Nevertheless, Kanban doesn't prescribe which models and methods *must* be applied. This is because the demands and situations are different in every organization. Neither does Kanban prescribe *how* models and methods should be applied. No assertions are made as to what is right and what is wrong. Just for the various types of visualization, there are already as many possibilities as there are organizations in the world that apply Kanban.

2.5 IMPLEMENTATION OF THE CORE PRACTICES IN AN ORGANIZATION

The "lack" of prescription is on the one hand a great freedom that Kanban allows in individual cases of application. This is because the existing processes in the system themselves *are* the basis for change resulting in improvement. Strict prescriptions would be counterintuitive, because situations in organizations are too multifaceted: the development of software for a web shop is guided by different laws than, for example, the development of a navigation system for an airliner. On the other hand, however, this lack of prescription is of course also a weakness of Kanban from the perspective of those seeking patent remedies and best practices. But this is how Kanban's core is defined:

> *Adopting Kanban doesn't mean adhering to predetermined policies. Kanban helps you understand the work relationships in your own organization and in this way supports context-specific learning. Your own personal reflection is expressly allowed!*

In Kanban, we have a diagnostic instrument that encourages us to change certain aspects of our situation by highlighting problems. It is not in itself a method for solving these problems—this requires the resourcefulness of the participants—because Kanban doesn't administer predefined solutions but rather promotes the development of a self-sufficient team or system toward a kaizen culture. The result is that the people in the system begin to reflect upon existing processes and experiment with them. In time, a constant workflow will evolve where even the iterations are surplus to requirement. The observable and measurable results of this workflow take the pressure off the individuals and make possible reliable assertions regarding the completion of tasks. This promotes trust between those paying for and those doing the work.

WHAT YOU CAN TAKE AWAY FROM THIS CHAPTER

The greatest corporate demand of our age is to raise productivity in knowledge work in order to remain competitive on the global market. The invisible processes of knowledge work however often make it hard to find the correct starting points for improvements. One knows that something isn't working as it should, but it's hard to identify exactly where the problem lies.

Kanban doesn't concern itself with the individual but rather with the system in the optimization of processes. Individuals can only work as well as the system in which they operate allows them to. The starting points for changes assisted by Kanban are always the current processes—no idealized system status is designed and implemented in advance. The basis is the visualization of the current workflow and the quantitative restriction of the number of tasks underway in a process in order to shorten the lead times. People therefore have the opportunity to improve processes themselves, at their own speed.

Kanban's goal is to develop a kaizen culture step-by-step, a culture that is focused on providing better results for the organization in economic and, for those working for the corporation, social terms.

Kanban's core practices are:

1. Make work visible.
2. Limit work in progress (WiP).
3. Manage flow.
4. Make policies explicit.
5. Implement feedback mechanisms.
6. Improve collaboratively (using methods and models).

Kanban makes very few prescriptions about *how* something should be done. Kanban makes suggestions *that* something should be done.

3

VISUALIZATION

The starting point for a Kanban implementation is consensus on the desire for change with the help of Kanban as well as consensus on the overriding goals of this change (we will go into further detail on this topic in Parts 2 and 3 of this book). This desire for change will ultimately not only result in an improvement of the working conditions of the employees but will also have a positive effect on the competitive ability of the organization in the medium and long term.

In his reflections on competitive advantages and competitive strategies, the management theorist Michael E. Porter drew on the concept of a value chain. A value chain encompasses all primary and support activities necessary for the production of a product, which will differ in nature for every organization. According to Porter, competitive advantages are to a large extent determined by how well or badly the activities in the individual stages of the value chain are managed. With Kanban, we approach a clearly defined section of the entire value chain in order to achieve improvements (and thus changes). Let us remember that Kanban is concerned with evolutionary change management, which means simply that we are constantly developing ourselves further through self-selected, small steps, each building on the previous steps. Kanban is initially concerned with the visualization of the workflow. It is in this way that the technical kanban system and its tools become the motor of change.

Kanban Change Leadership: Creating a Culture of Continuous Improvement,
First Edition. Klaus Leopold and Siegfried Kaltenecker.

3.1 FIRST STEP: DEFINING THE EXTENT

In theory, one could depict the entire value chain of an organization with a kanban system, which, in its way of thinking and its instruments, is designed in such a way as to be applicable to all fields. Kanban may have evolved due to reflection upon improvement in the field of software development, but it is actually flexible enough to not only work for software development. The application possibilities range from personal task management through execution of insurance claims to portfolio and program management. The one difference lies in the fact that different questions are to be answered for the different applications: in one case, it could be the very specific bottlenecks, while in another it could be the too high variability of processes.

Kanban can be scaled up step by step all the way to the point where it is applied to the entire organization or the entire value chain. Often, however, one makes the clear decision to start in easily observable surroundings, with one or more teams responsible for one part of the value chain. Observing and working on the entire value chain in one step could firstly be an arduous exercise and secondly be related to very large change, which is not what we are aiming for because in most cases it would immediately overburden the employees and, as in the present situation, be unnecessary. In the context of visualization of the workflow, "evolutionary" therefore means that we initially only examine a part of the value chain and start optimizing the processes there. Hence, we begin specifically with the part that the people who want to work with Kanban can also influence creatively.

Even before visualization has started, it must be decided which area (which section of the value chain) should begin with Kanban, for example, software development. The Kanban team comprises the people who work together in the chosen section of the value chain (e.g., development, testing, quality control, etc.). However, this team does not remain static during the further evolutionary development if at some point it becomes clear, due to the application of Kanban, that other constellations would be more productive. The Kanban team identifies the process and at which point of the value chain it is located by working and defining limits through the following questions: Which are its value-generating activities and which are someone else's? What sort is the sequence of the individual steps of the process? Where do their responsibilities overlap with those of other areas of the organization?

Establishing these boundaries also shows the intersections with other areas of the value chain and other units within the organization. The Kanban team agrees on policies for collaborative work with these upstream and downstream areas and units. We have often observed how in the course of events various smaller teams have melded into one larger group, thus giving themselves the capacity to react flexibly and the ability to reduce the number of blockers. Naturally, this also causes the boundaries—the points at which operations are taken or passed on—of the team to shift.

3.2 SECOND STEP: VISUALIZING THE PROCESS

Once the boundaries have been established, the next step is to make the operations and workflow visible with the assistance of a board or other aids. While identifying the individual steps of the process through which the workflow passes, Kanban groups should not let themselves be tempted to make the mistake of simply illustrating the official process as stipulated in project handbooks. Of course, there are organizations (such as military or infrastructure) that are required to adhere to strict processes. However, apart from these exceptions, official processes usually exhibit the weakness that they only exist on paper and barely correspond to actual reality. Such nonexistent processes are the wrong starting point for change. To orient ourselves around them would unnecessarily delay the change and/or improvement.

In a technical kanban system, it is always the process currently being used in real life that should be visualized.

The visualization is therefore also a task for the Kanban team—only the team knows how it actually functions.

The identified steps in the process are listed in columns according to their operational sequence. Figure 3.1 shows a sample workflow of analysis, development, and testing represented using a visual board. As with most things in Kanban, there is no recommended layout for the board. We have seen boards visualizing the workflow in spiral form and boards using a motorway as a metaphor—anything that expresses the process as sensibly and clearly as possible is permissible. Many teams explicitly take note of the completion criteria ("definition of done") for each step so that all team members share the same understanding of when the work has been finished.

Irrespective of which form is chosen, the board should always have two specific characteristics:

1. The input queue on the far left of the board is where the work item types that should be completed first are listed. These items are the result of the

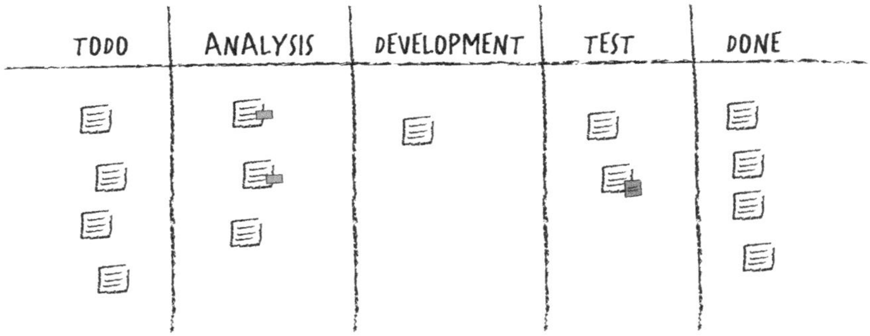

FIGURE 3.1 Example representation of a Kanban board.

conversations in the so-called queue replenishment meeting, which will be discussed at greater length in Chapter 6.

2. The point of handover in the final column on the right-hand side of the board—in this example entitled "Ready for release"—is the point at which the work leaves the kanban system (often simply indicating completion with the word "done"). A ticket landing in this column doesn't necessarily mean that a product is complete. In many cases, it simply signals that the next stage in the organizational value chain—the stage that borders with that section—can take over the work on the next part of the product.

MISUNDERSTANDINGS

Kanban and the waterfall

Kanban is a pull system. This means that "one piece of work" out of the whole group of work items is singled out, worked on, and then indicated as being ready for the next step in the process. This is sometimes falsely compared with the waterfall method in software development. This is a traditional way of doing things whereby *all* items that have been defined in advance for a particular step in the process must always be completed. Thereafter, the finished work items are passed on to the next step *all together*. However, this is precisely what Kanban does *not* do.

Kanban and creativity

Some people are critical of Kanban because for them the very creative act of developing software is not reducible to a process of steps that follow each other sequentially. They say that Kanban is therefore not applicable to creative software development. Regardless of whether they are creative or not, there are always activities that have to be carried out sequentially. For example, a software feature is first programmed and only then transferred onto a server, the same as always. One of our Kanban colleagues employs a very efficient trick in his workshops when someone in the room claims, yet again, that software development doesn't follow a sequence of steps. He turns it on its head and asks, "What do you do before the release? And before that? And before that?" He *always* gets answers to these questions. Then he reads the answers out in the correct order and asks, "And is that not a sequence? Why should that restrict your creativity?"

3.2.1 How Are Work Items Visualized?

The simplest way is to represent the individual operational work items with Post-it notes or cards that display what should be done (Fig. 3.2). We can very easily step into the currently running process by organizing the current to-do work items into the corresponding process steps as "tickets." The tickets are then moved into the column on the board in which work is currently taking place. And please note according to the pull principle and *not* the push principle!

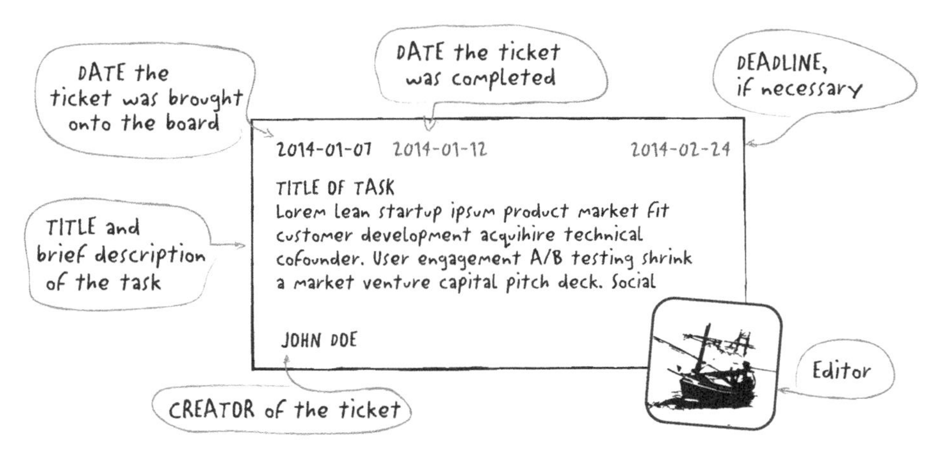

FIGURE 3.2 Example of a work item ticket.

In the principles of Kanban, we said that we also want to analyze our workflow in order to be able to identify problems. One of the most important metrics is lead time. If we take note of when the ticket goes onto the board, when it goes into a particular step of the process, and when the work item is complete, then we have already established information about lead times. Together with other information, this can give us a deeper insight into the factors disrupting the workflow.

A deadline has been noted on our example ticket. With Kanban, we essentially want to escape the single-minded world of deadlines. Rather, we want at some point in the future to be in a position where we are able to deliver reliable expected dates of completion to our clients, based on our observations and measurements. However, the immediate elimination of all deadlines is neither possible nor sensible for two reasons:

- Firstly, in fixed-date businesses, deadlines cannot simply be eliminated. In certain situations, software amendments are necessary at a specific date because, for example, a new law would be coming into force then. In Kanban, these types of jobs are also considered in classes of service, something we will be looking at later.
- Secondly, we want to respect the currently existing process. If work in this process is currently based on deadlines, then we'll leave it as is for the time being. As we progress with Kanban, deadlines will most likely become obsolete. In a flow-based system, as one gathers experience, defining projects specifically and being detached from the day-to-day become increasingly superfluous. They often become absorbed into the workflow. However, this development must be desired; it doesn't happen on its own and you must constantly stay on the ball.

In practice, it has proved to be useful to record the creator, that is, the "requester" of a task, on the ticket and to represent the person currently working on the ticket with

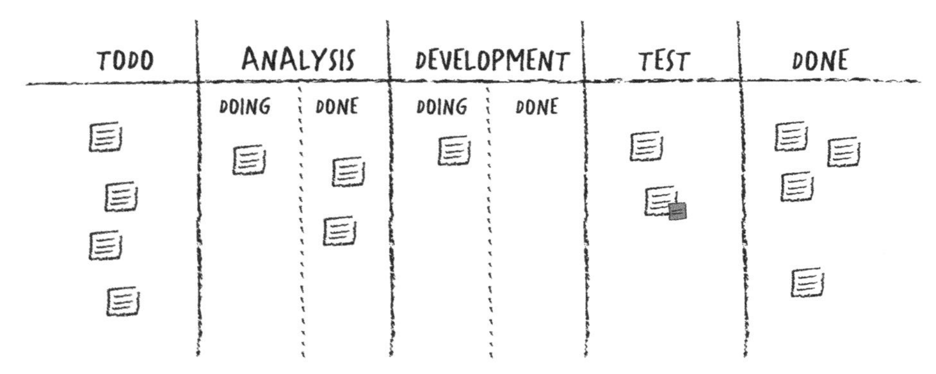

FIGURE 3.3 Kanban board with done queues.

an avatar. This makes further inquiry easier since the person currently working on the ticket isn't necessarily the person who can provide specific information about the entire process.

Work on a ticket can be interrupted due to any number of reasons related to missing information or missing infrastructure. In such a case, the ticket is simply marked as blocked, perhaps with a red sticker. A continuous workflow is one of the main goals in Kanban, and therefore, the removal of blockers interrupting the workflow should be a high priority.

As we now know, operations that are completed in Kanban are not pushed to the next process step; it is rather the worker in the next process step that pulls the work from his upstream colleague when he is ready for it. Completed work could be marked with a green sticker on the work item ticket or by separating the step's column into two subcolumns labeled, for example, "Doing" and "Done" (Fig. 3.3).

We will now introduce "queues." Incomplete work waits in a queue until it can be worked on further. The columns "Done" on our example board are such queues.

3.2.2 Representation of Parallel Processing

It is of course not always possible to carry out all steps one after the other in actual practice. Certain activities occur simultaneously. Parallel activities (parallel processing) can be represented on a Kanban board very well. Let us suppose that both the Development and Test Development steps occur simultaneously. We shall single out two of the many possible representations in order to demonstrate the type of effect different visualizations can have.

3.2.2.1 Variation 1 The simplest method is to record the two activities in the step "Development and Test Development" on one ticket as shown in Figure 3.4. Once both activities have been completed, the ticket moves over to the "Done" queue for this step of the process.

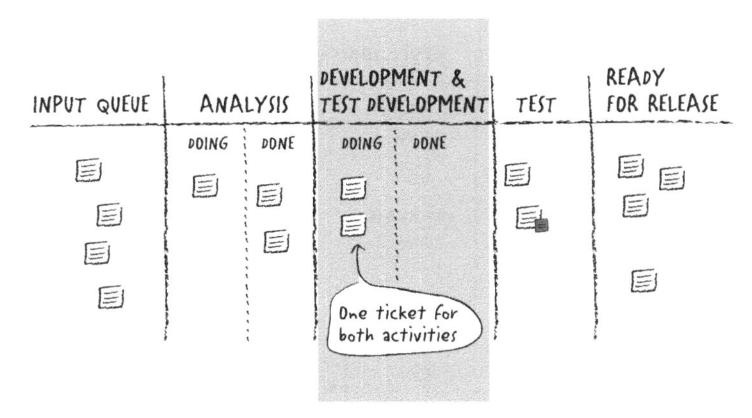

FIGURE 3.4 Representation of parallel operations using a ticket.

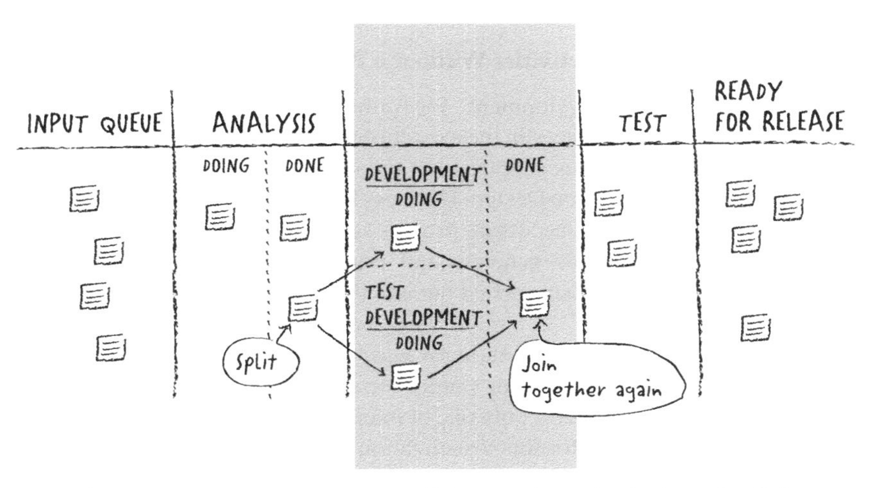

FIGURE 3.5 Representation of parallel processing using horizontal splitting.

3.2.2.2 Variation 2 The second method involves subdividing the step "Development and Test Development" visually into its two constituent parts and horizontally subdividing the column "Do" as shown in Figure 3.5. In the previous step of the process "Analysis," when the tickets are ready for further processing, they are simply copied and assigned to both Development and Test Development for their respective segments when they are ready for the work. The previously separated tickets are then joined back together again in the "Done" column of the step "Development and Test Development." Only when both tickets for the same work item are in the "Done" column can the downstream step "Test" take the ticket.

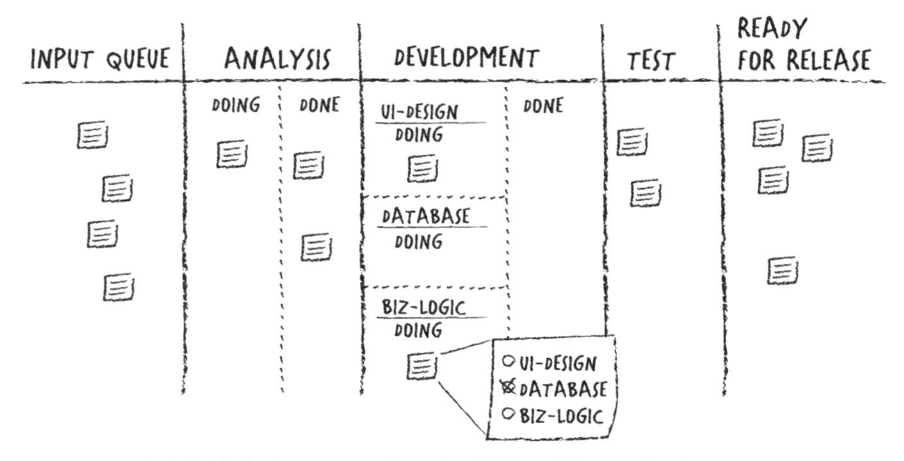

FIGURE 3.6 Representation of activities without a fixed sequence.

3.2.3 Representation of Activities Without a Fixed Sequence

Let us consider the step "Development" for further analysis. It could be the case in development that fixed sequences in the execution of tasks aren't always sensible, for any number of reasons. Let us suppose that the step "Development" is subdivided into the activities User Interface Design, Business Logic Programming, and Database Development. In this case, all necessary process steps are simply listed on one single ticket, especially when teams generate and print the tickets electronically. The individual process steps are then carried out and their boxes checked once they have been completed (Fig. 3.6).

This solution implies that all work items *must* be completed. If this is in fact not necessarily always the case, you can simply incorporate this by emphasizing what needs to be done. This is however only one of many possibilities. We will not attempt to list all possible variations for this visualization in this book; we will instead only present some tried and tested examples from real-world practice. What works in one situation might not work in another situation or in another organization. Ultimately, the visualization is concerned with dealing with the process and the work being done. Above all, it is about finding your own way and your own solutions and not using ready-made formulas.

3.3 DETERMINING THE WORK ITEM TYPES

One purpose of visualization in Kanban is to establish limits within the entire value chain of an organization. What we have seen so far in order to explain the principle of visualization is the visualization of a simple process in which only a small number of not particularly diverse operations are carried out. Of course, reality is not quite as straightforward. A team doesn't always carry out the same work item types, and a

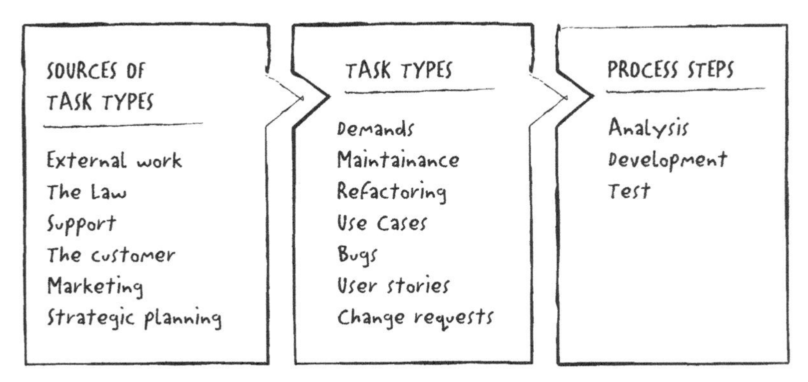

FIGURE 3.7 Derivation of work item types via the origin of the requests.

work item type isn't always processed through the same steps as those we've already identified. We thus need the flexibility to be able to represent such knowledge on our Kanban board visualization.

Real-world practice shows us that in order to answer the question "What work are we actually doing?" we often need to take a detour through the **origins of the work items**. We again require the boundaries of our Kanban system in order to do this: From whom does the team receive work items or inputs and to whom does it pass on the completed work items or outputs for further processing in another part of the value chain? Once the team has identified its process steps, it identifies its communication interfaces: Who needs something from us? Whom do we need to speak with? Who are our communication partners? By concerning itself with these questions, a team will for the first time become aware of the types of work—which **work item types**—it is completing in the individual **process steps** (Fig. 3.7). You frequently won't be able to identify all clients in one fell swoop right at the beginning of the Kanban implementation. "Aha moments" are often experienced in the course of implementation, when the team only gradually realizes what it is that it actually does and that the direct client often isn't the actual client. Business analysts, for example, often submit requests to the Kanban team; however, the actual client is the legislative authority because it was the one that decided to change certain legal precepts that must be fitted into software and structures. Such requests must most likely be treated in a different manner than routine requests. By constantly observing where work items come from, the Kanban team can pinpoint their system boundaries even more clearly. Ultimately, we're concerned with better understanding both the client and the business, thus constantly changing and improving the system so that it can respond successfully to even the most varied requests.

It is sensible to collect work item types together into groups for visualization purposes. The work item types "Bug," "Error in the live system," and "Configuration error," for example, are very similar to one another. One must ask the question of whether one wants to remain at this high level of granularity (and whether it is even necessary) or whether perhaps an overarching category "Bugs" would be sufficient. Work item types can be built from the most varied points of reference such as:

- Kind of work
- Origin of work
- Size of work
- Arrival rate of work

Since each grouping will disclose different "secrets" about the process during actual application, when you think of grouping, you should keep the following crucial questions in mind: What do I want to achieve with the visualization? What is the main goal of this Kanban initiative? What information do I need in order to achieve this goal? We're not interested in visualizing everything as precisely and fussily as possible. The idea is to visualize work item types *in a meaningful way*.

Grouping of work item types according to size—for the sake of simplicity, Kanban often uses the T-shirt sizes S, M, and L—could be very worthwhile if one of the important goals is due date performance to establish a reliable understanding with clients. So there are work items that can be completed in 1 h and others that require 20 days. With the size units for the work items, we can make reliable estimates of when we will be finished with a particular job because in this way we will shift our attention to the lead times and variability in our process.

3.3.1 Visualization of Work Item Types

There are also very different ways of going about representing work item types. Here, we present the most widely used visualization method, which is also one of the simplest: swim lanes. It is nothing more than a horizontal subdivision of our board. Let us suppose that we have four large groups of work item types that reach regularly across our process steps Analysis, Development, and Test. In our example, these work item types are Features, Change Requests, Bugs, and Support (Fig. 3.8).

All tickets relating to features are pulled through the process in the swim lane labeled "Features"; all tickets that relate to change requests follow the swim lane labeled "Change Requests"; and so on. The work item type "Support" is an exception since it doesn't follow the workflow of the other work item types. The insertion of such a "subboard" for a given type of work could have a number of reasons: it could be that support simply follows a completely different workflow or that the team knows that a ticket in this swim lane won't be on the board for any longer than a couple of hours, for example, and too much energy would therefore be expended constantly relocating the ticket to different columns. Hence, a simple separation into "Doing" and "Done" will suffice. The real value of the visualization here lies rather in simplification than in too-precise representation: the expenditure

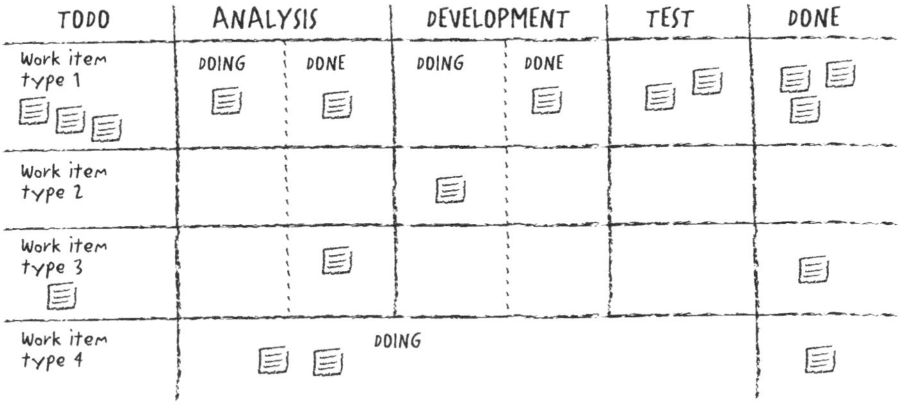

FIGURE 3.8 Kanban board with swim lanes for the individual work item types.

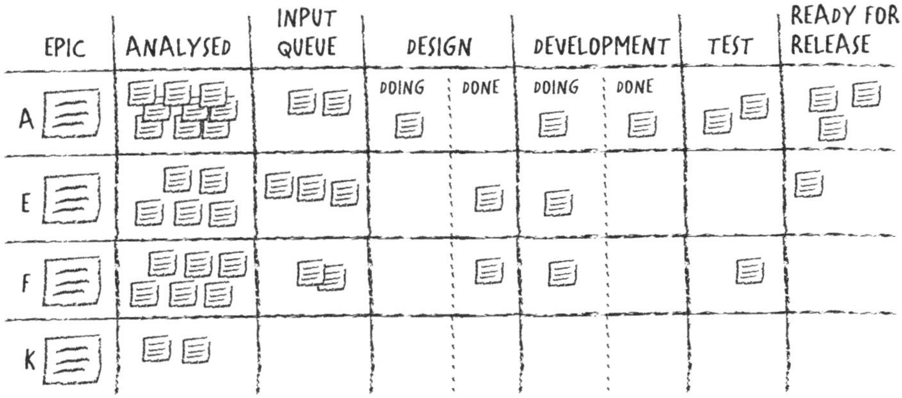

FIGURE 3.9 Splitting epics into smaller user stories.

of time due to unnecessary relocation of tickets can also be viewed as waste from a lean perspective.

Epics, all-embracing requests for the team, are very frequent occurrences in software development. They are large user stories that must be split up into many smaller user stories in order to be processed. Epics can also be visualized as individual swim lanes for each of the smaller user stories (Fig. 3.9). In our example implementation, this would mean that two columns are inserted on the left-hand side of the input queue: the description of the epic is listed in the first and the individual user stories in the second. Only when all user stories have moved across the board is the epic complete.

WHEN "OVERVISUALIZATION" IS USEFUL

There are teams that intentionally visualize everything that passes through the process on any given day very precisely. In one of our Kanban implementations, a team moved up to 80 tickets per day through the process. Each employee had a miniboard on their desk with four process steps across which they had distributed the Post-it notes displaying their tasks. The completed tickets were collected by the team members at the end of every day and hung on the main Kanban board. The employees wanted to know how much work they performed per day and how much of this was "invisible" work, that is, jobs that are completed within 5–10 min.

But the team also had another reason for this "overvisualization." They are in fact a support team, and the employees wanted to identify the causes for inquiries they thought had already been dealt with but that were nevertheless repeatedly resubmitted to them. The team clustered together similar tickets and set about seeking the reason for the repeatedly returning inquiries. Due to this detailed collection and clustering of the tickets, the team was finally able to identify a tiny misunderstanding in the user interface as the source of the problem. This small issue had caused many hours of additional work per week.

We have now established the boundaries of our Kanban team, identified process steps, and found out which work items we receive and from where. We are now ready to go ahead and distribute our capacities among the individual work item types. The workflow can now be controlled purposefully because behind every work item type there is a particular need and urgency. Bugs, for example, must be processed immediately, and features need a high degree of reliability in their delivery. The analysis of the input and the work item types also makes it possible to account for dwell times used for quick responses to unpredicted results or regular "bursts" (e.g., customers' wishes for specific changes). Our next task is the definition of WiP limits and classes of service for our Kanban system.

WHAT YOU CAN TAKE AWAY FROM THIS CHAPTER

Kanban achieves improvements (and thus changes) in a clearly defined section of the overall value chain of an organization, more specifically in precisely the section we can directly influence. The start and end points of the workflow in this section must therefore be established. The initial Kanban team asks itself questions regarding which tasks are theirs and which aren't, how the individual process steps in the workflow are ordered together, at which point operations are commissioned, and where finished work is passed on to a downstream stage for that particular section.

Collaborative policies are to be agreed upon with upstream and downstream interfaces (stakeholders) at the boundaries of the Kanban team. The process steps of the workflow of the team are most often visualized in columns on a board, electronic or physical. The team thus analyses and visualizes the *actual process currently happening*. The individual work items move from left to right across the board on Post-it notes or cards. These tickets contain important information for measurements such as start and completion time frames, process duration in the individual steps, or causes for blockers, for example.

There is a limitless variety of representations for processes in Kanban. Parallel processes and operations without a fixed sequence can be represented easily and without difficulty. What we are ultimately concerned with in the visualization is an examination of the process and the work to be done.

The team additionally identifies the work item types it repeatedly processes and examines the origins of these tasks. This ultimately helps assign the work items the appropriate importance in business terms, in the form of classes of service.

Different work item types can be visualized on the board with swim lanes.

4

WiP LIMITS

It's a very seductive premise indeed: the employees can surely do this one job quickly and should be able to manage that one too, and they're bound to manage to fit in this, that, and the other few work items somewhere along the line as well. But the problem is that in an organization there normally isn't only one rather spontaneous client; there frequently are many. Clients who themselves normally work to other people's demands—such as external customers—or have become victims of their own time management fall into this category. You can fill a funnel to the brim but that doesn't mean more water will flow through it. Quite the opposite: there's the danger that some will spill over, to be lost and forgotten.

In our consultancy practice, this sometimes leads to curious results. One of the teams we worked with revived work that had lain dormant in their backlog for 3 years, although it had been assigned a "high priority." This is the fate of many work items that are submitted to a team: after having been considered urgent at the start, they lie for months or years on end without anyone giving a hoot about them. In order to spare the development or Kanban team's nerves, and above all to lead an organization with good business sense, such operating styles should be avoided at all costs; they are a particularly frequent form of waste. In Chapter 2, we established that it is far more economically expedient to bring one work item to 100% completion than ten work items each to 10%.

Kanban therefore establishes the so-called work-in-progress (WiP) limits. These restrict the number of work items in the Kanban system as a whole in order to avoid

Kanban Change Leadership: Creating a Culture of Continuous Improvement,
First Edition. Klaus Leopold and Siegfried Kaltenecker.

the overflow described earlier. This means that parallel operations are neither impossible in Kanban nor undesired. A faster, more continuous workflow is achieved through this limitation, reducing lead times because team members do not need to continuously change from one work item to another. It works very simply in practice: write the number of operations allowed above every process step visualized on the board in both the "doing" as well as in the "done" columns.

When we talk of limiting the WiP, we are thus talking about restricting the number of work items allowed to be processed at that stage of the work.

The WiP limits are established in the course of designing the Kanban system. Working with Kanban, it will become clear how and where WiP limits must be changed. WiP limits are not set arbitrarily but on the basis of empirical experience. However, we're not concerned with a conclusion that is established once and then carved in stone. Just as with the whole Kanban system, WiP limits are objects of constant change and will always readjust to new circumstances.

4.1 THE ADVANTAGES OF WiP LIMITS

Continuous improvement is central to setting WiP limits, but the aims of keeping tied-up capital at a minimum and being able to honor commitments to customers are just as important. The motto for economically expedient work is: stop starting; start finishing.

Besides having a soothing effect on the nerves of developers and customers, what are the advantages of WiP limits?

- **Avoid switching between work items**: If you set yourself and others limits for the quantity of work you produce, you avoid the temptation to constantly switch between work items. This is why nothing gets finished sooner, and instead, everything gets finished later. WiP limits help to remove the inventory of accumulated unfinished work.
- **Lower lead times**: If we think back to Figure 2.1, it becomes clear how parallel operations have a negative effect on the lead times of each of the work items involved. Although we can prove this fact mathematically, we can also establish it using pure logic: since the artificial bending of time is unfortunately still the realm of science fiction, we human beings only have access to a finite amount of time for work. Doing more work within the same amount of time always comes at a cost to someone or something else. Normally, it affects product quality or employee health—most often both. If the number of simultaneous operations is limited, people can better concentrate on single work items. The better their concentration, the faster is the work item finished, and the sooner is the customer satisfied. One of the most important goals of WiP limits is therefore to keep a process flowing. Three parameters govern the flow of work items through a queuing system [1]:

- The average number of work items in the queuing system (L)
- The average waiting time for a work item in this system (W)
- The average arrival rate of work items (λ)

The average number of work items in a system is a product of the average arrival rate of work items and the average waiting time in the system:

$$L = \lambda W \tag{4.1}$$

Hopp and Spearman applied this principle to operations management and established the connection between throughput (TH, the average output of a production system per unit of time), WiP (the sum of products between the start and end points of a production process), and lead or cycle time (CT, the average time from the job release to the end of the production process) [1]:

$$\mathrm{TH} = \frac{\mathrm{WiP}}{\mathrm{CT}} \tag{4.2}$$

This very simple comparison, *Little's Law*, makes it clear that the waiting time in the system (the lead time) declines if the arrival rate of work items decreases.

- **Higher quality**: This is a direct result of the possibility to concentrate better on fewer work items using WiP limits. Diligence decreases as the number of parallel work items increases. We additionally reduce feedback times using WiP limits because downstream process steps can provide continuous responses since we are easily able to survey the group of work items being performed. Without WiP limits, developers usually receive such feedback about programming errors much later. The later you receive feedback about errors in your work, the more time you spend trying to identify these errors rather than fixing them. In a system with a continuous workflow, such responses are instantaneous; hence, the elimination of errors only costs the workflow a small amount of time.
- **Greater degree of predictability/due date performance**: If we are constantly working on many work items simultaneously, we lose sight of the amount of time we actually need to complete particular work items independently of the others. We ultimately constantly think about the simultaneity of the work items and calculate all sorts of temporal security buffers in order to make ends meet. Let us assume that a developer begins work on a project he knows will last 5 days. New work items continuously appear and he constantly ends up switching work items and consequently won't be able to honor his commitment to complete the original project in 5 days. WiP limits help a team identify how much time their work items actually require in order to produce concrete delivery date estimates and stick to them.
- **Fewer disruptions**: As we will be seeing, WiP limits are based on agreements. Stakeholders must also accept WiP limits, which will result in a change in their

way of thinking. People will think more fundamentally about whether the small favor they're asking is really so important or whether it isn't perhaps more sensible—both financially and socially—to let the team work in peace, free to concentrate on its current project. The process of change here begins to exceed the boundaries of the actual Kanban team because the stakeholders start considering what is really necessary and what is in fact waste.

DO WiP YOURSELF

Before reading any further, go ahead and see for yourself the sort of effect various WiP limits would have on a workflow.

Starting point

- A Kanban team carries out three working steps (A, B, C).
- One member per step is a specialist for that step.
- We want each Kanban team member to be working on at most two work items at any given time.

Build a Kanban board that illustrates this situation and simulate the workflow with yellow Post-it notes. Bear in mind that work items can also be blocked.

Questions

1. What situations could cause the workflow to falter?
2. What would be the effect of reducing the WiP limit to one?
3. What would be the effect of increasing the WiP limit to four?
4. What effect do the various WiP limits have on the input queue?
5. Suppose work proceeds more slowly in step B than in steps A and C:
   a. How does this affect the whole system?
   b. As a manager, how would you proceed?

We shall now look at two very significant advantages of WiP limits: how they make blocker-related problems and bottlenecks visible.

4.1.1 Making Problems Visible

When jams in certain stages stop work in upstream or downstream stages because operations are constantly marked with red blocker stickers, you can specifically research the causes: Where do the constant blockers come from? First of all, a team should dedicate itself to solving this problem before starting new work. Should this not happen, it's very likely that the problem will repeatedly manifest itself at other stages, usually a couple of magnitudes larger than originally.

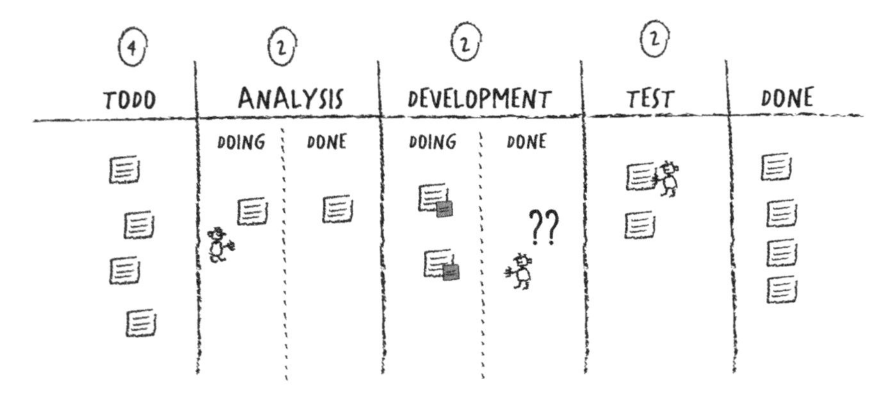

FIGURE 4.1 Idle loops affected by blockers make problems visible.

In Figure 4.1, no additional work can be taken on in the stage "Development." The WiP limit of two is—for whatever reason—already fully occupied with the blocked operations. This naturally has an effect on the up- and downstream stages. In a (pull) system that is not run with WiP limits, a developer can take the finished work from the "Done" column in the "Analysis" stage. This option is no longer available with a WiP limit of two; the developer is not *allowed* to begin any new work. We have thus identified a problem and we should "stop starting and start finishing." We first eliminate the blocker in order to complete the operations.

Were the employee to ignore the WiP limits and constantly take on new work items leaving the two operations blocked, the following would happen: as soon as the blockers are removed and the developer can work further on these two work items, all the other work begun in the meantime is left unattended. Additionally of course, the lead time of the blocked work increases. Without WiP limits, the problems would only be deferred, not eliminated. Ideally, the developer who cannot proceed due to the blocker should dedicate this free capacity to analyzing the causes of the blocker, asking, "Why have the blockers evolved and how can we avoid them in future?"

4.1.2 Making Bottlenecks Visible

We can identify where the bottlenecks in our workflow are through a combination of visualization and WiP limits. Employees working on the bottleneck cannot collect work from upstream colleagues because they are still occupied with the current work item. And since the upstream colleagues also stick to the WiP limits, they cannot pull any new work into their stage of the process either. The workflow will gradually grind to a halt. These bottlenecks are the actual motors of improvement because we seek the reasons for these bottlenecks—just like with the blockers—and eliminate or widen them.

Let us look more closely at the situation in Figure 4.2. The employee in Development has completed two work items and thus transfers them to the "Done" column. Work in Analysis has also been completed, but the developer is not allowed

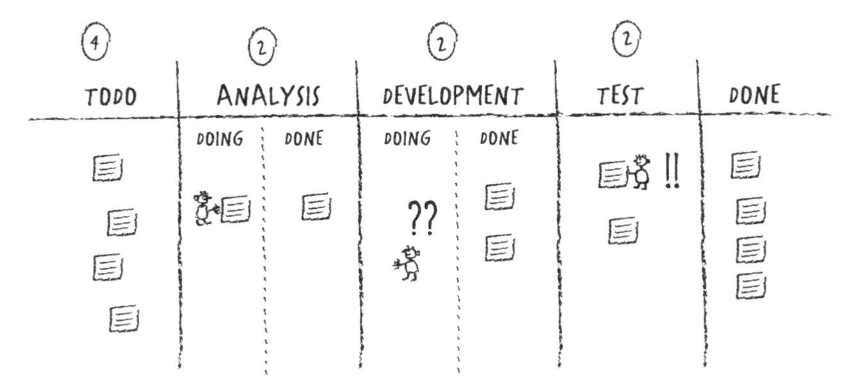

FIGURE 4.2 WiP limits make the bottleneck in the stage "test" visible.

to collect it because he observes the WiP limit of two. Test turns out to be the bottleneck in this process because the Test employee is still working on his current work item, observing the WiP limit of one. Only when the tester has completed his work and gathered a new work item from Development can the developer begin with the next work item.

The entirely logical solution for the provisional elimination of the bottleneck is to have the developer assist the tester with his work, which is however easier said than done. The major objection is that developers are employed to develop and not to test. A second objection, more of an emotional nature, could be that the developer doesn't see himself as a tester or might have reservations for technical reasons. Many teams make life easier for themselves by assigning risk profiles to operations that colleagues from other areas are allowed to carry out. Thus, everyone is clear on what a developer can or cannot do in the testing process.

Traditional management generally views a currently "unemployed" employee as someone being paid to do nothing. A shift in perspective would reveal that this currently unemployed employee has the time *right now* to focus on improvements. It is therefore valuable to have an employee simply available. The entire professional groups are in principle paid to wait around or simply for their availability. Just think of firefighters: many hours pass by without any demand for their skills and the group occupies itself, for example, with ensuring its kit is in good working order. However, when a call comes in, they are all instantly available. If help is needed in other areas on a phase basis, Kanban explicitly welcomes this "slack" as a source of process improvements and as "support contingent." Quick reaction needs slack. Classical project management is constantly obsessed with the idea of using all employees at full capacity, if possible 100% of the time (or more). In our example in Figure 4.2, more and more work would simply queue up in Development if we were to follow this way of thinking. However, if we consider the theory of constraints again (see excursus), then it is significantly more expedient to raise the throughput of the system rather than running at full capacity until breakdown.

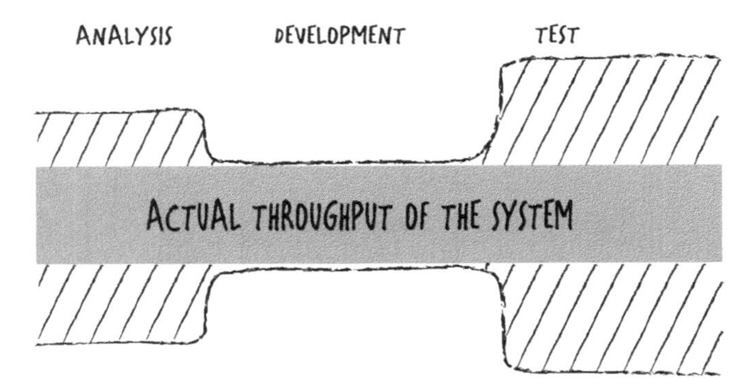

FIGURE 4.3 Bottlenecks produce "slack" that is useful for improvements.

To understand this, let us look once more at a flow-based system but in a different way (Fig. 4.3). The system consists of the stages Analysis, Development, and Test, whose height in the representation corresponds to the number of work items that can flow through the individual stages. Quite clearly, the number of workable items in Development is significantly lower. Since Kanban and the theory of constraints state that no more work items than the bottleneck can cope with should be allowed into the system, we tailor the WiP limits in such a way that the bottlenecks are not overburdened. A case in point is the turnstile, which is an artificially generated bottleneck in museums and libraries, for example, that makes the fact that you are a hare or a tortoise completely irrelevant because you can pass through no faster than the turnstile allows. The turnstile determines how many people pass through in a given time (= throughput). It is exactly the same in work systems: the bottleneck determines how many work items can be completed in a given time. Therefore, we should only let as much work into the system as the bottlenecks can cope with. In so doing, we free up work and thinking power not being used in other stages to help out at the bottleneck and furthermore to widen it. The promotion of cross-functionality in a team—consisting of knowledge sharing among team members and investment in training so that employees can work flexibly in various areas—can be a fixed goal. The 100% use of the capacities of employees not working on the bottleneck isn't much use because the bottleneck is still there and the potential for improvement and innovation remains untapped. If we use this naturally occurring potential, organizations wouldn't need expensive, artificially generated improvement and innovation programs. This is often the hardest aspect of Kanban thinking for management.

EXCURSUS: THEORY OF CONSTRAINTS

The physicist Eliyahu M. Goldratt, who later became a management advisor, made a very insightful observation: the main woe of many organizations is their fragmentation. Structuring an organization as many small units is supported by the belief that complexity is better controlled in this way and that the optimization of every unit will have an automatically optimizing effect on the whole organization.

In most cases, the hoped-for success will never be achieved despite exploring all options. In his theory of constraints, Goldratt postulated that it is usually never more than *one single* significant bottleneck that prevents a system from achieving its goals. In its performance capabilities, every system, regardless of whether it's concerned with production or knowledge work, is automatically defined by this bottleneck. Just as a mountaineering team can only ever climb as quickly as their slowest member, the throughput of an organization and its subgroups is determined by the bottlenecks.

> *Our goal shouldn't be to use our resources—in this case, people—at an ever-greater capacity. The goal must be to improve the throughput of work.*

The five steps for the elimination of bottlenecks and the increase in throughput suggested by Goldratt are nothing other than a continuous process of improvement [2]:

1. Identify the bottleneck.
2. Decide how the bottleneck can be used most effectively.
3. Make everything else subordinate to the above decision.
4. Widen the bottleneck.
5. If the bottleneck is eliminated, start the process again.

Goldratt strongly warns against being satisfied with tackling the bottleneck and eliminating it. There is the danger that you focus too much on the current hurdle and forget that an organization consists of interlinking, interdependent units, hence the iterative approach. We will again encounter the theory of constraints as a tool for improvement in Chapter 7.

4.1.2.1 Dealing with Bottlenecks In Kanban, we deal with bottlenecks by buffering them so as to ensure that the employees in the bottleneck always have enough work to do and that the bottleneck never runs dry should upstream-process employees go on holiday or fall ill. The buffer evens out variability by raising the WiP limit for the entire system. In the example illustrated in Figure 4.4, the process step Test is the bottleneck; therefore, a buffer "ready for test" has been inserted. Now, if a developer goes on holiday or if there is any other variability, this will be absorbed by the buffer. The developers still continue to work with a WiP limit of two, but they put their completed tickets into the Test buffer, where the testers also collect their work items. We have therefore not changed the WiP limits of the individual stages, but rather raised the overall WiP limit from five to eight. Buffers should be as small as possible but large enough, however, to establish an even workflow.

4.1.2.2 Moving Bottlenecks It is often difficult to tell the difference between a real and an assumed bottleneck. A real bottleneck in a system exists if capacities,

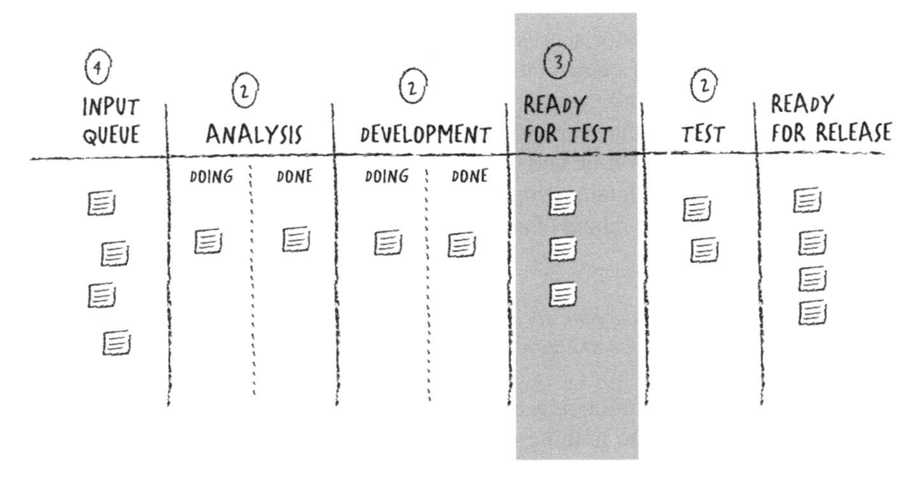

FIGURE 4.4 Buffering a bottleneck.

> **BUFFER OR QUEUE?**
>
> They may have a similar function but they aren't actually the same thing. Let us think back to Chapter 3: incomplete work waits in a queue to be further worked on. Each "done" column within a process step is one of these queues.
>
> Likewise, a buffer is a special type of queue for incomplete work. However, buffers are an "intentional" design element in particular situations with the scope of establishing and maintaining a constant workflow. Buffers thus serve to cope with an accumulation for further processing by creating an "interim storage" for the work. They are intentionally inserted in order to absorb irregularities such as variations in the speed of work. A well-applied buffer can "save" lead times in this way. In visual terms, a buffer looks like an additional column on the board.

resources, or abilities are consistently overburdened at a particular point or step in the process. In order to recognize this, the workflow must be observed over a reasonable period of time. It is only then that one notices that there are also bottlenecks that only appear for a small amount of time. In knowledge work, this is the rule rather than the exception.

Let us look at an example of impractical requests. Suppose a business analyst wants to make a particular algorithm faster by a couple of milliseconds. The analyst probably needs about 5 min in order to process this request. However, the development team is for the next few weeks occupied with the sole task of searching through the entire code for a potential place where 2 ms could be saved, and to alter it accordingly. In the workflow, this development would now obviously appear as a bottleneck, although in actual fact it is not. In another situation, the business analyst has the task of replacing an existing recommendation algorithm in the software (think, e.g., of Amazon) with a new one. He requires many weeks for the development of the new formula, which the developers however write into the code very quickly: old formula out, new one in. Done! In this example, the development is supposedly the bottleneck and, a little later, the business analysis. However, this is only due to the variability of work items. In these situations, to adapt the process immediately and invest in additional resources in order to widen the bottleneck in the long term would constitute a bad financial decision.

The purpose of visualization and WiP limits is therefore not to search for instant solutions to all potential bottlenecks but rather to discover bottlenecks, observe them for a period of time, understand them, and then carry out the appropriate changes if necessary.

4.1.2.3 The Collection of Work A specific form of bottleneck is the specializations of individual employees. It is in many organizations the case that one employee possesses very detailed know-how in a particular field and can thus support many teams or departments. Since the specialist continually wanders through the value chain and is not constantly available, the work that he comes to inspect must be ready for him at the right time. Simultaneously, this dependency must not be allowed to affect the continuous workflow in other stages of the process. Hence, we insert another buffer for the completed work items to be collected in the intervening period.

The collection of work items in a buffer is also a good solution for stages of work in which the sequential processing of tickets would create an artificial bottleneck (Fig. 4.5). "Test" is in many cases a candidate for a buffer: unless the tests are highly automated (e.g., with manual acceptance tests), a tester often has the same expenditure regardless of whether he or she must test 1 ticket or 10.

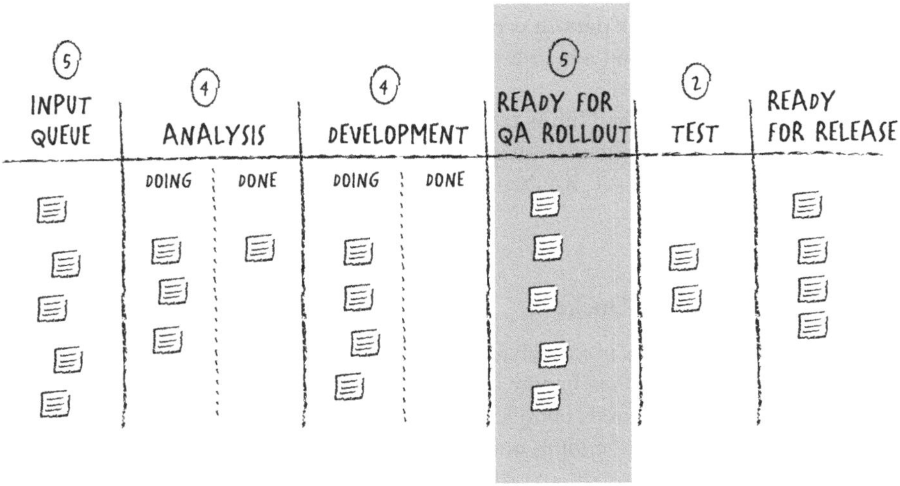

FIGURE 4.5 The collection of tickets.

4.2 SETTING WiP LIMITS

One of the most frequent questions in Kanban is: "What is the *right* WiP limit?" There is no single correct answer to this question; no magic formula. The only way to find the right WiP limits for a particular Kanban system is trial, observation, and modification. In Kanban, there is no way around this empirical approach.

However, there are of course thought-based aids that can help you in your quest for the ideal WiP limit. In our work with Kanban teams, one particular formula that considers the number of people involved was very beneficial for us. For example, let's say we know that there are four people working on one process step. Let's ask them the question of how many work items each person can sensibly process simultaneously. If we establish that two work items per person can be processed simultaneously, we simply multiply the four people by two and establish a WiP limit of eight. If a new employee joins or if one leaves the team, we don't need to establish a new WiP limit. The formula allows the limit to be scaled up or down very simply based on the conditions.

The formula, of course, could also be: divide the number of employees by two. Pair Programming is a typical application for this. We intentionally allow less work into the system than can be handled by employees because the employees must work together. This is a fundamental requirement for Pair Programming, but you can of course support the exchange of knowledge in other fields in this way if, for example, it is a stated goal of a Kanban implementation.

Three general statements can be made from our observations:

1. A **WiP limit < 1 per person** makes sense if multiple people should work on a single work item.
2. A **WiP limit = 1 per person** is often a good solution in "mature" organizations with few blockers and little variability.
3. A **WiP limit > 1 per person** is most often helpful when the process is affected by many blockers and a high level of variability.

We can also separate the WiP limits from the people and ask: "How can we apply the limits so that the waiting times for work in queues are reduced?" However, particularly in the introduction of Kanban, the person-centered approach is easier to understand.

4.2.1 Size of the Input Queue

The number of work items in a Kanban system is limited using WiP limits. The input queue on the board is the clear handover point for jobs submitted to the Kanban team. Anything in the input queue is completed first. The lead time clock begins to tick as soon as a work item is in the input queue, and the Kanban team can make promises concerning lead time. However, in order to have any chance of keeping these promises, the size of the input queue must be limited: clients should only be allowed to hand over a certain number of jobs to the team such that the input queue also has a WiP limit. **What** lands in the input queue is determined at the "queue replenishment meeting" (see Chapter 6). **How large** the WiP limit for the input queue depends on the throughput of the team and the frequency of the replenishment meetings.

SIZING: AN EXAMPLE

Suppose the replenishment meetings in a Kanban system take place on a weekly basis. Our team manages to action 10 tickets per week on average. The deviation is between +4 tickets and −2 tickets. Thus, practically speaking, the team actions between 8 and 14 tickets per week.

An expedient WiP limit for the input queue is therefore 14. If the WiP limit were to be set at 10, this could cause the team to experience idle moments due to the deviation.

If we scale up our example to a level of general validity, we can say the following:

- **A small input queue** results in smaller lead times. Here, a team can react in an agile manner: since the input queue empties itself again quickly, new work items can ideally be picked from the backlog before the next replenishment meeting, and the team can address customer needs earlier.
- **A large input queue** results in longer lead times—because more work items must pass through the Kanban system—and decreased flexibility. It is often the case that organizational financial priorities suddenly change in the course of doing the work. The higher the input queue's WiP limit, the longer one has to wait until there is a free slot for the job and the possibility of assigning new priorities for the jobs. Hence, it is only much later that we can react to changes in the situation.

4.2.2 WiP Limits for Various Work Item Types

In Chapter 3, we demonstrated that even the most varied work item types a team has to execute can be represented most easily using swim lanes. This subdivision, in combination with the WiP limits for the respective work item types, makes it possible to control the workflow as required and minimize risk. Figure 4.6 shows that a WiP limit of eight has been established for the work item type "Features," that is, a maximum of eight Feature Tickets are allowed in the system (from Analysis to Test) at any given time. The same is true for the other work item types: a maximum of two Change Requests, a maximum of three Bug Fixes, and a maximum of two Support Tickets can swim through the process.

The stipulated WiP limits in the swim lanes indicate the emphasis of the Kanban team. In Figure 4.6, the team's current focus is on Feature Programming, and therefore, the WiP limit for Features has been jacked up. If it were to concentrate on

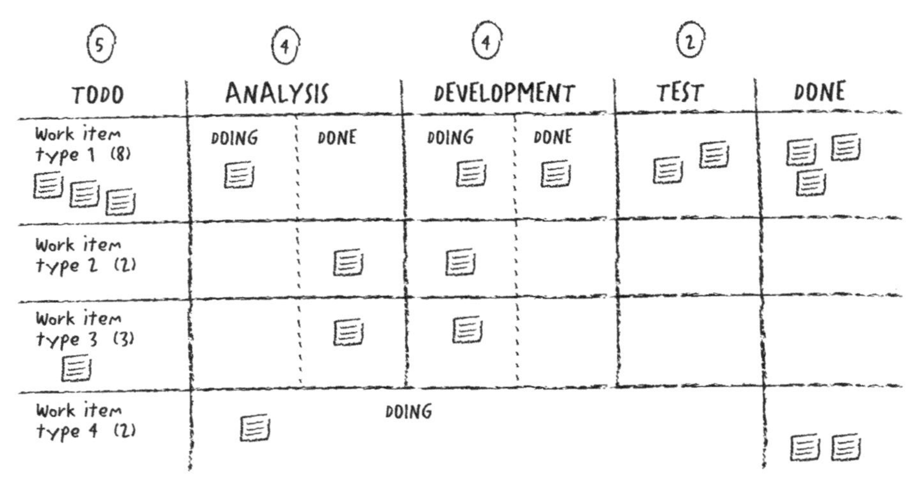

FIGURE 4.6 Control of the workflow using WiP limits for work item types.

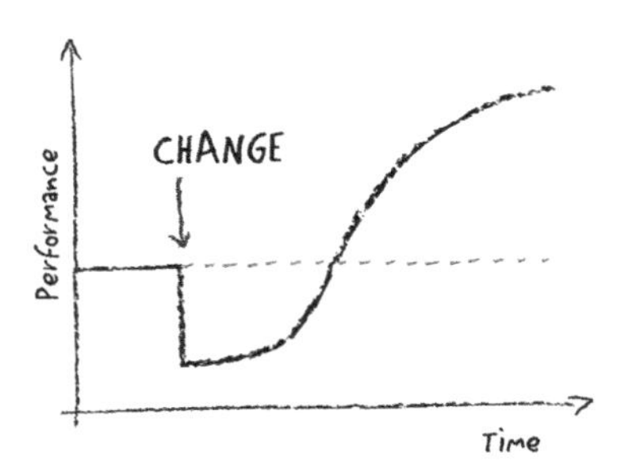

FIGURE 4.7 J curve effect.

customer satisfaction, the team would alternatively adjust the WiP limits to favor Change Requests or Bug Fixes.

4.2.3 Consequences of Different WiP Limits

We don't use Kanban just for the sake of using it; we always set our sights on particular goals. There will be consequences for the workflow as well as the mental state of the organization—or the part of the organization we're currently considering—and its possibilities for further development depending on the WiP limits chosen. These consequences can be helpful but can also hinder if not dealt with sensibly. One can observe the so-called J curve effect (Fig. 4.7) with every change, regardless of whether it is due to Kanban. The performance of a team drops at the beginning of a change but in most cases subsequently increases in the fullness of time to a level above the original one.

The level of the WiP limits influences how drastically the performance of a team drops at the beginning of a change.

- **Extreme case 1—WiP limits are too low**. If the limits are set too low, many problems become visible all at once. This causes trouble because the team is occupied solely with the task of eliminating the problems rather than completing the actual work. On the one hand, resistance develops because whoever is suddenly confronted with such a multitude of blockers and problems very quickly seeks refuge in the comfort of the "good old days." On the other hand, the team also notices that it works better once it has eliminated the problems. But realizing this and subsequently achieving the increase in performance take significantly longer. The point is that the problems are already there before the implementation of Kanban, but with Kanban, they become visible.
- **Extreme case 2—no WiP limits**. Even with difficult starting points such as high market pressure, it's useless to set no limits at all since neither problems nor bottlenecks would become visible. In the best-case scenario, we would only be able to see small improvements on our J curve, even after much time has passed.
- **Extreme case 3—WiP limits are too high**. If the boundaries are extended upward, work items can still go through the system, even though many blocked work items are hanging there, stuck. Problems do indeed become visible, but they are ignored because everyone is sufficiently occupied and still has work to do. In such a situation, thinking about improvements is simply not an option.

As far as WiP limits are concerned, only in extreme cases can we with a great degree of certainty say that they will be disadvantageous. As to which WiP limit is right or wrong in any particular case, each of you must find that out yourself. The most important thing is to constantly check the feasibility of WiP limits and adapt them. Of course, that doesn't mean you should change them at every opportunity and at every meeting, but rather when they are no longer helping you achieve your stipulated goals at their current level.

One thing is required in order for WiP limits to work: consensus between the team, management, and stakeholders. Only when all are in agreement that they are to work with WiP limits, what these limits are, and what they communicate can constant transgression of the boundaries be avoided. The stakeholders do not need to understand the mode of operation right down to the last detail, but they do need to understand the advantages of working with WiP limits. It becomes particularly clear when a stakeholder himself stands in front of the board and sees how additional work affects the lead times. Right at the beginning, the initial level of the WiP limit is a matter of gut feeling and only of secondary importance. What *is* important is the agreement that work is being done in a WiP limit-based pull system because that is the motor of change. The more realistic tailoring of the WiP limits to the needs of the customers will take place as the work progresses.

WHAT YOU CAN TAKE AWAY FROM THIS CHAPTER

In order for a team to be able to honor commitments for work completion, it must limit the number of work items that pass through the process simultaneously. In Kanban, we use WiP limits in order to achieve this. By limiting the number of work items that are processed simultaneously, we prevent costly constant switching between work items, lower the lead times, increase the quality via short feedback loops, and improve the due date performance.

The most important effect of WiP limits in terms of improvement is making visible the blocker-related problems in the process and the actual and shifting bottlenecks conditioned by variability. Bottlenecks produce slack (freeing up employees) that can be used for effective process optimization. One can deal with bottlenecks and variability by introducing buffers in the Kanban system to provide optimal efficiency at the affected points and smoothen out the workflow. It is important however that one distinguishes between temporary and actual bottlenecks and only takes corresponding measures after a certain amount of observation.

In Kanban, the input queue also has a WiP limit. How high this limit depends on the throughput of the team and the frequency of the replenishment meetings. In order to control priorities, the individual work item types can also be assigned WiP limits.

There is no such thing as a generally valid recommendation for a "right" WiP limit. Most of the time, you approach the appropriate WiP limits for the process of a team through trial, error, observation, and further trial. The most important criteria however are consensus between the team, management, and other stakeholders for the use of WiP limits and that these boundaries are respected.

5

CLASSES OF SERVICE

2011 was for the world of IT security what Queen Elizabeth II would probably call an "annus horribilis"—a terrible year. It started with WikiLeaks publishing decrypted diplomatic cables, but the knock-on effect this had was significantly worse for organizations worldwide. Hackers such as Anonymous don't just want to make shady political events accessible to all—they also highlight uncomfortable data protection issues. Suddenly, every organization and every authority, regardless of whether a giant company or government organization, had to assume that confidential customer data could be dragged into the public domain. This of course is the worst-case scenario. Such a situation is not about resolving a technical security problem. With such actions, an organization incurs costs in the form of loss of trust and damage to its reputation, which are rather high considering the expensive communication campaigns required as remedy.

In such dramatic cases, it is clear to all members of an IT department that they must drop everything in order to resolve this acute problem as quickly as possible. Fortunately, "near catastrophes" are the exception in everyday life, but they make it particularly clear that the work items an IT team has to process out differ in the impact they have upon the business. Thus, it is expedient to differentiate between work according to type and extent of the consequences. We are confronted with different performance, services, and treatment everywhere. On the one hand, it is a strategy for the maximization of price elasticity. For example, in air travel, certain target customers are prepared to pay more for better service and more room between

Kanban Change Leadership: Creating a Culture of Continuous Improvement,
First Edition. Klaus Leopold and Siegfried Kaltenecker.

the seats. The airline therefore treats them differently for the duration of the interaction. However, differentiation is also necessary to control complexity in work processes and, for example, make routes easier to plan. Think of parcel services. These organizations differentiate between express and normal delivery and the routes of the drivers are planned accordingly. The differentiation of work items is also the norm in IT: bugs are classified according to their impact and those with a high impact are eliminated immediately, with all necessary resources being concentrated on this one work item. In many cases, there are even special contingency plans for emergency fixes, patches, or releases.

Kanban uses **classes of service** to sort work items according to impact and risk and thereby be capable of reacting at the right moment at a reasonable cost. Classes of service also help make reliable assertions to stakeholders about which services can be delivered within a specific time frame (**service-level agreements (SLAs)**).

The categorization of work items with classes of service means treating these work items differently in order to better control their lead times and to allocate capacities accordingly. Classes of service are also a tool for increasing customer satisfaction.

WHY NOT "PRIORITIES"?

You might be asking why Kanban uses the concept of classes of service rather than simply describing it as a "prioritization" of the work items. The problem with prioritization in software development is that it is often forced upon developers by people outside actual development. Most teams are prevented from acquiring a better understanding of these issues because they lack the actual (business) risk information that lies behind them. It is this information that would, for example, absolutely justify a priority of level one. Priorities are so-called proxy variables because, although one can objectively measure their results, an objective reference basis is lacking. In many cases, the reference basis is the personal pressure on a product or project manager by the customer. If decisions concerning priority are taken on this very biased basis, every work item would suddenly have a priority of one ad infinitum.

5.1 COST OF DELAY AND POLICIES

Do you remember the team in Chapter 4 that revived a 3-year-old work item from its backlog? This work item was apparently not so important from a business perspective, or else someone would have enquired about it earlier. Unfortunately, it's simply not the case that all work items are quite so irrelevant. Had someone accompanied the work item with relevant risk information 3 years before, it would have perhaps been immediately clear that it wasn't necessary to begin working on it right away since, in this case, inactivity wouldn't have had any noticeable consequences. In the case of many products and projects, an organization doesn't sense any psychological strain for a long time unless they are immediately initiated. However, at some particular point in time, everyone's suddenly called to arms. In the case of yet other work item types, everyone's immediately in the blocks purely as a matter of principle, for example, when the pay function doesn't work anymore on an online shop. Classes of service are designed to make these objective consequences visible to all and efficient risk management possible. Therefore, classes of service as opposed to priorities are always based on objectively comprehensible risk information: either commercial information about the economic impact should a project either not be completed or be completed later than planned (cost of delay) or concrete policies about how the individual classes of service must be treated (policies).

The costs of delay that form the basis of our classes of service are a function of the business impact we can visualize as time progresses. We also derive several policies for every class of service from them. Similar to the checklists in the cockpit of a plane, these policies help us to establish the appropriate processing order in various situations and thereby minimize risks and are of course also subject to continuous improvement. How classes of service are defined and what names they receive should be decided specifically by each organization. Common classes are, for example:

- "Expedited"
- "Fixed delivery date"
- "Standard"
- "Intangible"

Simply taking predefined classes of service out of books, however, is not expedient because time–impact relationships are always different in every organization. The basis for classes of service is unique to every organization: they are either based on the organization's own measurements from the past or on professional estimations. The thrilling thing about classes of service is that everyone concerned becomes aware of their impact and must get involved with them. In development teams, even in the case of the most die-hard technicians, a greater corporate awareness evolves because of them. The team's immediate stakeholders are normally present for the discussions in which the classes of service are defined. They are closer to the market and know exactly what sort of external impact certain things will have. Therefore, there must be agreement with them with respect to classes of service and their different treatment.

The **number of different classes of service** and the respective policies cannot of course be infinite. In order for the principle of classes of service to work in practice, each employee must know the classes like the back of their hands and be able to relate the correct policy structure with them. Employees must *understand* these policies and the reasoning behind them. In practice, it's proved worthwhile to work with a maximum of six classes of service, although it is fundamentally a case of finding a balance between them being easily memorable and being able to react flexibly to customer needs. The following also holds for the visualization of classes of service:

the simpler, the better. Colors still have the best ability to convey meaning, and with their help, most people will intuitively figure out what is meant. The easiest solution therefore is to represent the different classes of service using differently colored tickets. Other types of tickets, adhesive panels on the tickets, or individual swim lanes are sometimes also used.

5.1.1 The Class of Service "Expedited"

All projects that immediately and directly incur high costs are assigned the class of service "expedited" (Fig. 5.1): an attack by Anonymous, for example, or the crashing of crucial servers in the case of a web product. There can also be positive reasons to expedite cases, such as a product feature that opens up new areas of business for an organization if it is released very quickly. For a "quick win" to be possible, the team must quickly step into the breach. If an expedited ticket arrives on the board, the WiP limit can be exceeded. What the prioritized treatment of these tickets normally looks like (e.g., which capacities are made available for them) is established by the policy structure the team decides on. On the graph, the dotted line symbolizes the processing time, while the solid line represents the impact on the business.

The policies for "expedited" tickets might be as follows:

- An expedited ticket must be collected as soon as an appropriate employee is available; other work items must wait.
- The WiP limit for any stage in the process can be exceeded.
- Only one expedited ticket may be processed at any one time: the WiP limit for this class is one.
- The completed ticket is immediately deployed.

Effects: Since an expedited ticket is processed significantly more quickly, its lead time decreases. Of course, this comes at a cost to other work currently in progress: their lead times increase as do the lead times for other expedited tickets potentially being processed. Thus, when we don't limit the number of expedited tickets in the

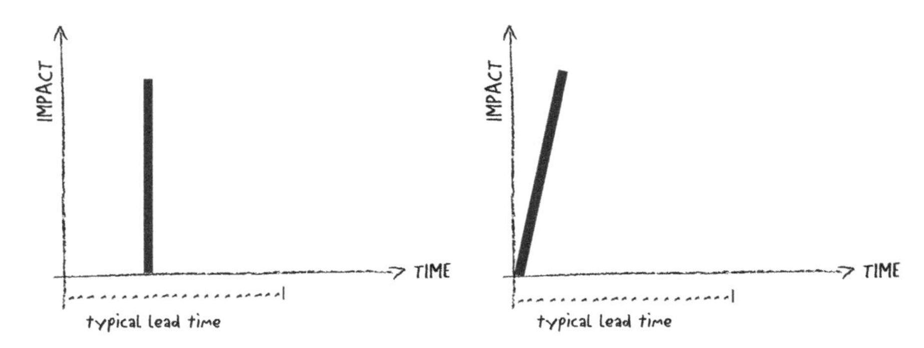

FIGURE 5.1 Progression of costs of delay for the class of service "expedited."

system, they no longer perform their function. Again, this is precisely where we notice the difference between Kanban and systems that work with priorities: without objective information, for example, cost-of-delay graphs, everything is important and everything should be done immediately. If we don't impose any limits on ourselves, we block the entire system and increase the lead times. This becomes a vicious circle. If work with a priority of one is not completed quickly enough, stakeholders fear that their jobs that have been assigned a priority of two would slip off the radar. So they assign everything a priority of one just to be on the safe side and burden the system even further because all of a sudden the team is solely occupied with completing expedited tickets. Therefore, the following must be observed: *if you allow expedited tickets, their number must be restricted.*

5.1.2 The Class of Service "Fixed Delivery Date"

In an organization, expedited tickets should be the exception rather than the rule. It is far more often the case that development teams are confronted with fixed delivery dates. In this case, the team has a specific time window available during which to complete the work. If the work item isn't completed by then, there is the danger that the organization could suffer high costs or be impacted badly, either with immediate effect or after a period of time (Fig. 5.2). An example of this is a change in legal requirements. Such changes come into effect on a specific date. If databases, payment gateways, etc. haven't been adapted accordingly, then the organization risks incurring fines, complaints from customers, and, in the worst-case scenario, a total inability to do business.

Tickets with a fixed delivery date are thus not dealt with immediately but with a higher degree of attention.

The policies for "fixed delivery date" tickets might be as follows:

- An analysis is carried out to estimate the size and expenditure of these work items in order to estimate the lead time.

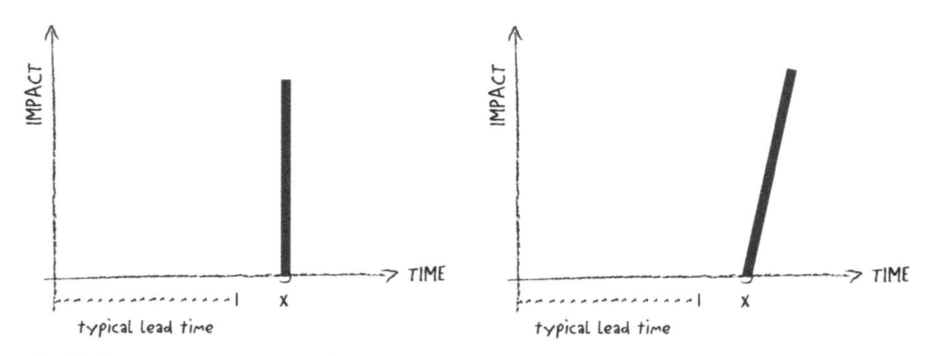

FIGURE 5.2 Progression of costs of delay for the class of service "fixed delivery date."

- Based on this estimate, work is begun early enough to be completed punctually, before the deadline.
- Work belonging to the class of service "fixed delivery date" adheres to the agreed WiP limits.
- If a project becomes so delayed that punctual delivery is at risk, the class of service can be changed to "expedited."
- Completed work items are put in the next scheduled release.

Effect: The positive consequence of this class of service is that work is guaranteed to be completed within the stipulated time frame. However, just as with expedited tickets, be cautious: in this class of service, work items are also processed with a certain priority, increasing the lead times of other work items as a result. Therefore, it is also the case that the number of "fixed delivery date" tickets in the system be limited. And everyone should be particularly clear about one thing: *desired* dates are not *actual* dates!

HOW IMPORTANT IS THE "FIXED DELIVERY DATE" CLASS OF SERVICE?

In our day and age, the obsession with deadlines can often be nerve-wracking. However, what we want to express using the class of service "fixed delivery date" with its basis cost-of-delay graph isn't necessarily quite the same thing.

For many software development teams, the reality is that the costs of delay are less important to the client than being safe in the knowledge that the problem will be processed and not constantly relegated down the list. That's why they set a specific completion date in their contract, mostly because they're not able to observe (and thus make forecasts about) the system that the team applies. As we all know from our daily work, requirements for deadlines evolve in many cases from mistrust.

In work systems not limited by WiP limits, new work constantly comes bursting into the system. As a consequence, teams end up planning based solely on deadlines and not on the basis of a workflow that they know intimately enough to be able to make reliable forecasts. But that's precisely what we want to achieve with Kanban.

Whether the "fixed delivery date" class of service makes sense or not depends on the context in which the team is operating. Support teams barely need these classes of service if at all, because most of the time they must simply complete their work items as quickly as possible. It's a little different for development teams in project environments, especially if the projects have a seasonal aspect. Platforms for the Christmas season in January would be a joke; they should of course be completed before the beginning of the shopping season if possible. In this field, fixed delivery dates are unavoidable and correspond to the definition of a class of service in its real sense.

Before applying these classes of service right at the beginning of your Kanban system, ask yourself the following questions:

- What is the environment in which the team is working?
- Are the fixed deadlines for the team's projects of such a nature that not meeting them will result in significant disadvantages? Or are stakeholders erring on the side of caution because no reliable forecast of when results could be expected could so far be made?

5.1.3 The Class of Service "Standard"

Unless we're in a highly chaotic or notoriously understaffed organization, "standard" tickets should form the bulk of our work. The related cost-of-delay graph indicates that with this class of service, we would not be threatened with costs immediately or after a particular deadline (Fig. 5.3). It is rather that we are here concerned with opportunity costs: the impact on the business presents itself in the form of lost profits. Suppose we decide not to apply a new function yet. With this function, we might have already been able to increase our turnover or competitive abilities as compared with other organizations.

Not getting these work items done immediately doesn't have a huge direct impact on the business. However, never getting them done will probably detract from the organization's competitive abilities.

Policies for the "standard" class of service might be as follows:

- Standard tickets are pulled through the system based on the first in, first out (FIFO) approach.
- Standard tickets are processed when no expedited or fixed delivery date tickets are awaiting processing.

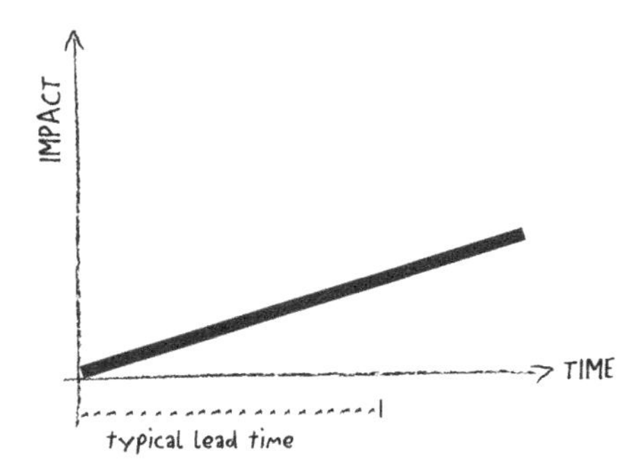

FIGURE 5.3 Progression of costs of delay for the class of service "standard."

- Completed tickets are included in the next scheduled release.
- No estimates with regard to lead times and effort are made.
- Work items can be analyzed in order to establish a sizing system, where S, *small*, could mean within 1 day; M, *medium*, within 1 week; and L, *large*, more than a week (note: these definitions would naturally vary from one organization to another).
- *Large* tickets must be broken down into smaller work items.
- Ninety percent of all tickets are to be completed within 10 days (note: here, we are already in the realm of SLAs, which will be discussed more thoroughly in Section 5.3).

5.1.4 The Class of Service "Intangible"

One knows that some work items could have an impact in the distant future. In many cases, no one knows exactly when this will happen. These work items are currently commercially important but not urgent. An example is a version upgrade of a software component. Initially, the upgrade might be ignorable because the software functions without a hitch in the whole system. However, as soon as dependent software components are no longer compatible with the old version, this has huge consequences on the entire system. It could thus be the case that the "intangible" work items are eventually promoted to a higher class of service and proceed to exert pressure, both commercial and in terms of deadlines (Fig. 5.4). However, we want to avoid this class of service evolution, and to this end, such tickets are to be constantly processed.

Policies for the "intangible" class of service might be as follows:

- Tickets can be processed by any team member as long as no ticket of a higher class of service (standard, fixed delivery date, or expedited) is in the queue.
- Expenditures and lead times are not estimated.
- The work items can be analyzed in order to establish a sizing system—large work items are further broken down into smaller tickets.

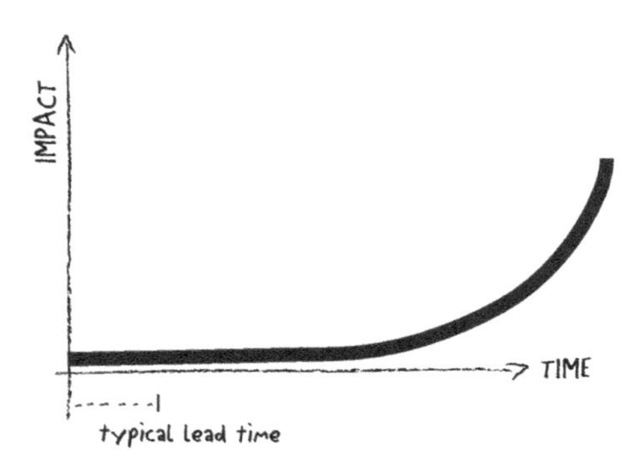

FIGURE 5.4 Progression of costs of delay for the class of service "intangible."

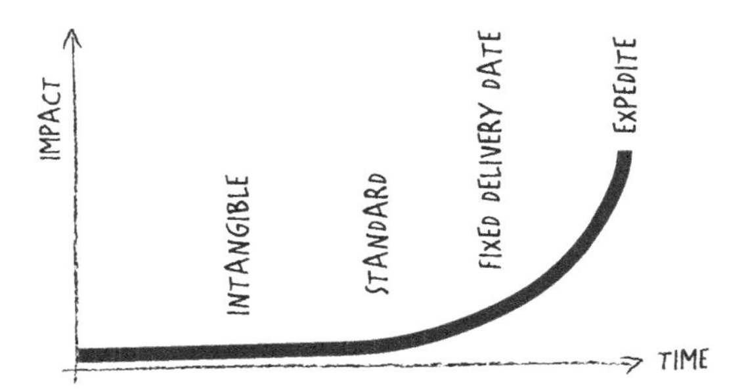

FIGURE 5.5 Change in the urgency of classes of service as time progresses.

- If a ticket with the class of service "expedited" comes into the system, the work being done on "intangible" tickets is paused.
- There are no SLAs for this class of service.

The indeterminacy of such work items shouldn't be taken as a sign that they are completely ignorable. Their sense and purpose is for work to begin early on, that is, when time is not yet pressing.

The work is important but not urgent. The reasons this class of service is so important despite its unsuspecting name are twofold:

1. We are here concerned with the kind of further development for which we never really have time. The idea behind these classes of service is to reserve capacities for these work items. We therefore might have a policy that there must always be at least two "intangible" tickets in the system. This class of service thus also serves as an engine for improvement and innovation that we integrate into our system.
2. This class of service is intended to protect teams from unpleasant surprises. If you can't afford to pay any attention to the work items with intangible costs and if you don't have enough of them constantly in the system, you will be able to see how the urgency of these work items changes drastically (Fig. 5.5): they are initially repeatedly delayed until suddenly the deadlines loom or external pressure is applied to expedite the work and pass it on for release. As time progresses, the ticket acquires an increasingly higher class of service until it finally arrives in the "expedited" class, thus having a large impact on the business. Therefore, always working on a certain number of "intangible" tickets is an advance time investment.

5.2 CAPACITIES OF CLASSES OF SERVICE

So how many tickets of each class of service should there be in a Kanban system? There is *no single right* answer to this question. David J. Anderson says, "The tasks should flow through the system in such a way that commercial value and customer orientation are optimized and the releases lead to maximum customer satisfaction." We have already seen that the number of tickets in the classes of service "expedited" and "fixed delivery date" should be, for understandable reasons, severely restricted. However, there should be quantitative limits for all classes of service in order to be able to control the workflow efficiently. Capacities and policies ensure that tickets receive the appropriate treatment based on their class of service. With a WiP limit of 20, for example, the following can be established:

- Five percent expedited = 1 ticket.
- Twenty percent fixed delivery date = 4 tickets.
- Fifty percent standard = 10 tickets.
- Thirty percent intangible = 6 tickets.

Why have we distributed 105% here? We previously introduced the policy that tickets with the class of service "expedited" can exceed the WiP limit and that only one expedited ticket may be processed at any given time. With a WiP limit of 20 for the whole board, the WiP limit is raised to 21 with the expedited ticket, equivalent to 105% if a WiP of 20 is our basis. The distribution of capacity across the individual classes of service in each actual case is a question of continuous trial and adaptation.

Capacities must be allocated if the idea of classes of service is to function because, depending on *how* the capacities are distributed, this will have an impact on the lead times of the individual classes of service.

5.3 SLAs

All steps so far—visualization, setting WiP limits, and ordering work items into classes of service—are intended to contribute to the optimization of lead times in the workflow. These improvements can only be achieved if processes are measured (we'll discuss this thoroughly in Chapter 7). A team must know, at any given time, where it stands in terms of performance capability. Only thus is it able to decide what its goals are and how it's going to achieve them. But for the Kanban team, measurements don't just have the self-oriented goal of improving things. The appropriate metrics help define the time required for any given work item of any size and with any degree of difficulty. This means that teams are constantly able to make more precise forecasts, based on the measurements, as to when a customer will receive the completed product.

SLAs are fixed commitments for the delivery of work items based on measurements.

This has two significant advantages:

1. You don't need to carry out any costly estimates based on trial and error.
2. You don't need to bother with the sorts of practices that are built upon a lack of trust (e.g., negotiated commitments).

SLAs can be made for classes of service, all sorts of sizing systems within classes of service, work item types, or combinations thereof. What might such an SLA look like?

- Eighty percent of standard tickets are completed within 14 days.
- Ninety percent of *small* standard tickets are completed within 5 days.
- Ninety percent of *medium* standard tickets in the "Feature" work item type are completed within 15 days.

WHAT YOU CAN TAKE AWAY FROM THIS CHAPTER

It is possible to distinguish between the work items a team has to complete in terms of business impact. It is expedient to differentiate these work items in terms of this impact and treat them differently according to their urgency. In Kanban, we use **classes of service** to do this. With time and ongoing measurement and observation, the team can make reliable forecasts of what can be provided within a specific time frame (**service-level agreements (SLAs)**) to stakeholders.

Classes of service are based on objectively comprehensible risk information about the economic impact of a work item should it not be completed or be completed with a delay (**cost of delay**). Usage policies for every class of service are stipulated according to the respective cost-of-delay graphs. The names of the individual classes of service and the risk information provided vary from one organization to the next. Frequently used classes of service are:

- **Expedited**: work that directly and immediately causes high costs. The number of expedited tickets in the process should be strictly limited.
- **Fixed delivery date**: enough time is available for completion of the work item. However, if the work is not completed before the agreed deadline, large costs with immediate effect or after a period of time will threaten the organization. The number of such tickets in the process is also to be limited.
- **Standard**: tickets of this class should constitute the bulk of the work. Delays in completion have no direct impact but can have an effect on customer satisfaction and market position.
- **Intangible**: are requests that are important but not urgent. These are primarily investments in future success.

Quantitative limits within the stipulated overall WiP limits are distributed among the individual classes of service. The allocation of the limits will have an impact on the lead times of the individual classes.

Using the lead times, a team can ultimately agree upon SLAs with stakeholders. These are fixed commitments for the delivery of work items based on measurements.

The target lead times agreed with the customers in SLAs are based upon past data, data often not yet available to a Kanban team just starting out. Therefore, teams must initially work with forecasts that can be adapted as soon as hard data is available. This data comes in the form of measurements of lead times, for example, and the spectral analysis of the lead times, which we will look at in greater detail in Chapter 7.

6

OPERATION AND COORDINATION

Up until this point, you've read a lot about the instruments that help a Kanban team visualize and control its workflow and shorten its lead times. Now, of course, you're probably asking how the team coordinates with its members and others. How does work reach the team in a Kanban system? Who decides which work items land in the input queue? Who decides what goes into the next release? And how?

Kanban is pure communication. We visualize the work and pull work across a board—even this itself is a type of communication. But even if, or indeed because, we make lots of things visible, this approach fundamentally obliges us to do one thing only: talk to each other. Anywhere people identify blockers, bottlenecks, or other problems using the visualization, they begin to discuss things and look for solutions together. Kanban is thus sustained to a very large extent by communication between team members. Admittedly, we're usually only focusing on one section of the entire value chain, but every Kanban team should be perfectly clear about the fact that Kanban is not a tool for self-gain; rather, it is the entire organization that should profit from Kanban. Together with the conversations that are anyway generated within the team, this functions above all through communication with the up- and downstream stages at the boundaries of the team's section of the value chain. The institutionalized form of communication for the operation and coordination of a Kanban system is thus meetings.

In the same way that Kanban prescribes very little apart from a few principles, it also doesn't dictate *that* meetings must happen, *what* kind of meetings they should

Kanban Change Leadership: Creating a Culture of Continuous Improvement,
First Edition. Klaus Leopold and Siegfried Kaltenecker.

be, or *in what manner* these meetings should take place. In everyday practice, the only thing that has proved to be of importance is that during the evolutionary steps of change there is always the need for periodical discussions and more intensive communication with the other stakeholders of the current operation. The smallest change that we would advise an organization to implement in this area is the daily stand-up meeting. For some incomprehensible reason, in knowledge work, there are few or no daily morning discussions for the coordination of the day's work. This is not the case in production (or, e.g., in hospitals), where it is and always has been customary to begin the workday or the shift by having the employees inform each other of work items to be done. In doing so, these organizations avoid needless duplicate and superfluous work that is both wasteful and expensive.

Together with the daily stand-up meeting, we suggest a few other meetings that have proved to be useful for Kanban operations, although they are by no means mandatory:

- Queue replenishment meetings
- Release planning meetings
- Team retrospectives
- Operations reviews

We want to establish a specific delivery cadence in Kanban operations. We also want to avoid waste—and time is that which is most often wasted. In order to minimize coordination costs, meetings should take place at regular intervals, always in the same place and at the same time.

6.1 DAILY STAND-UP MEETING

In the daily stand-up meeting, the team discusses and organizes the work items for the day, analyzes blockers, and looks for ways to remove them. Visualization using the board usually shows the workflow situation very precisely. The team meets at the board every day at a fixed time and works through the work items systematically. Although we read from left to right, teams often instinctively begin the discussion with the work in the far-right column on the board, that is, with those that are nearly completed—the influence of the feeling that it is better to have first completed one work item before beginning the next is here apparent. Particular attention is given to work that hasn't progressed in a long time or has been visibly marked as "blocked." It often happens, mostly with mature teams, that the blocked tickets are discussed almost exclusively. The teams discuss what the causes are and which employee at which point of the workflow can help. Similarly, anyone needing help can submit their request. In the context of the daily stand-up meetings, the team can clarify whether all tickets are progressing within the agreed service-level agreements or whether the class of service of individual tickets needs to be changed.

During the daily stand-up meetings, Kanban teams don't focus on the individual employees but rather on the *work* that must be done. In Kanban, it is not only at the daily stand-up meetings where it is forbidden to assign blame, it is forbidden in all types of meetings. In order for the daily stand-up meetings to run as effectively and efficiently as possible, deeper or more technical discussions between individual employees are siphoned off into follow-up meetings. Traditional meetings are often simply planned time wasting because a handful of colleagues discuss things with each other and 20 other people are forced to sit there and listen although it has nothing to do with them. This is avoided through follow-up meetings. The daily stand-up meeting may be the Kanban team's "own special meeting," but other stakeholders can of course also take part if interested or if the team so desires.

6.2 QUEUE REPLENISHMENT MEETING

How do the tickets get into the input queue? Via decisions on the order of the work items taken in the queue replenishment meeting. This meeting occurs regularly but can also be requested if desired. We are here already at the boundaries of the Kanban team's jurisdiction, and therefore, those present at the replenishment meetings should be the following:

- All those who assign work to the team
- All those who receive completed work from the team
- All those who are able to contribute to decisions regarding the team's next work items

Together with the team representatives who provide the technical aspect, internal stakeholders and representatives of external customers also participate and compete for the capacities of the team. Ideally, a senior member of management is there as a helper to contribute in terms of the entire organization's perspective. Ultimately, the places in the input queue are stipulated by the WiP limits; the decision as to when a work item is to be processed must also be made keeping business relevance in mind. It could happen that a change at the higher corporate levels of the organization occurs as a result of these meetings. After all, people meeting here have intimate knowledge of the strategic and tactical impact of various sequences of work items. Beyond these Kanban meetings, these people are very often disconnected when working. They come together in the queue replenishment meeting and, most often after an initial battle for the best place in the input queue, seek solutions that make sense for the entire organization.

In the classical approach, the team receives inputs from all involved in the ongoing work process. As we have already established, in such a situation, everything normally has a high priority since the individual wishes of colleagues aren't always based on considerations about costs of delay. Thus, in Kanban, we want to channel jobs in meetings, thereby avoiding the situation where employees are constantly dragged away from their current work items and have additional ones that may or may not be economically viable piled upon them, forcing them to constantly switch

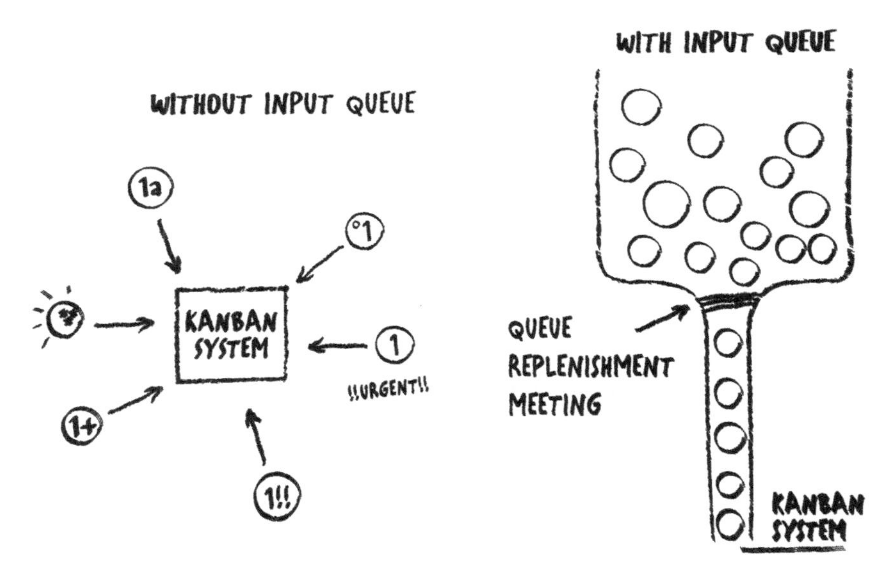

FIGURE 6.1 Channeling work items at the queue replenishment meeting.

between them. To do this, all work items are collected in a funnel during the queue replenishment meeting including, for example, a backlog, a ticketing system, or a request tool, *before* they are then put into the input queue (Fig. 6.1). We gather the affected stakeholders at one table so that they can *together*, rather than each for themselves, decide the order in which work items are to be passed on to the team and the classes of service these work items are to be assigned. The effect is that the team orders its work items, works in peace, and can better meet its deadlines, which is in the interest of the stakeholders.

DON'T CROSS THE LINE

In the case of one client, the team had already acquainted itself very well with Kanban. The queue replenishment meeting was already established but old habits die hard. Many of the employees had previously simply brought additional work items along to the individual developers. Although these employees knew that there were now different policies, they still stubbornly came to the development team's desks to give them work items that in their eyes were obviously urgent.

In such cases, highly visible measures are sometimes required. The team moved the room around; barricade tape was streamed in front of the work area, complete with the warning sign: "Before you cross this line, ask yourself whether your request wouldn't be better taken up in the queue replenishment meeting."

What is the optimal frequency for the queue replenishment meeting? In Chapter 4, we established that the optimal size of the input queue depends on the throughput of the team and the frequency of the queue replenishment meetings. Conversely, the meeting frequency depends on the throughput of the team and the size of the input queue. A weekly meeting is appropriate in most cases, but we have also worked with teams who would meet up at short notice if needed. There is no best solution. The ideal solution is a result of trial, observation, and change.

6.2.1 Backlog Maintenance

With the WiP-limited input queues, classes of service, and queue replenishment meetings, control mechanisms are created that should induce the stakeholders to have a very detailed discussion in terms of what they want the team to do for them. Of course, there is considerable danger that these wishes—should they fall victim to commercially more important work items—nevertheless land in the backlog and sit there, hoping for better times. However, it's pretty common that these work items are simply forgotten by the clients because ultimately everything wasn't quite so important after all. It's obviously more pleasant for a development team if they don't have to go around carrying a sack full of such work items. Additionally, participants in the queue replenishment meeting shouldn't have to work through a pile of work items time and again in order to establish which are to be put into the input queue first. For the sake of efficiency, the backlog should only consist of work items that have a real chance of implementation. It is thus advantageous for the participants of the queue replenishment meeting to examine the backlog closely at regular intervals and decide which work items could once again be removed. In most cases, you can establish policies for this, for example, "All backlog items that haven't been taken up in 6 months are to be discarded." In extremely dynamic environments, we have sometimes seen this deadline shortened to as little as 2 months. Other teams have opted for a two-stage process and give the ticket holder an initial warning that their item is on the "to chuck" list, allowing the former client to consider whether the ticket is still relevant or not.

Regardless of the approach:

A small backlog makes it easier to sort work into a processing order.

If, contrary to expectation, a work item that has been thrown away must indeed be implemented, then it makes its way back into the backlog and is brought up again at subsequent queue replenishment meetings.

6.3 RELEASE PLANNING MEETINGS

Unlike the daily stand-ups and queue replenishment meetings, the frequency of the release planning meetings depends on the point in time at which a release is to take place. Kanban doesn't have any requirements for the interval between releases. The delivery interval always depends on the context in which we're working and the quality or level of refinement of the product to be provided. However, certain

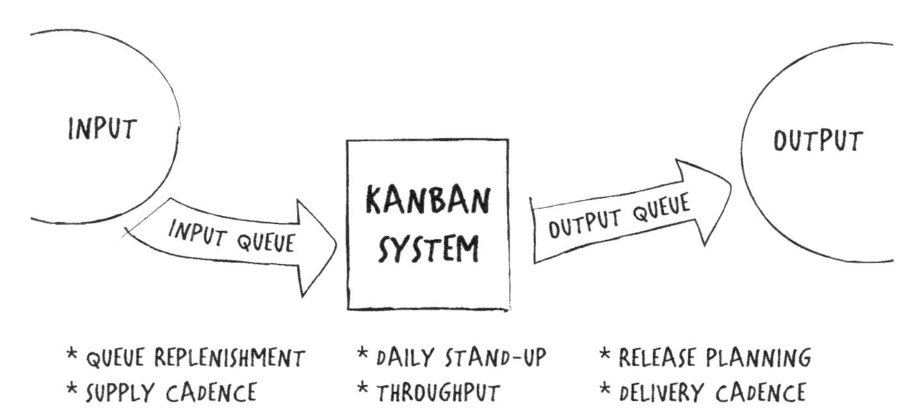

FIGURE 6.2 Relationship between the queue replenishment meeting, the daily stand-up meeting, and the release planning meeting.

regularity in delivery is desirable for the simple reason that it strengthens customers' trust. This is as true for software development as it is for public transport services. If you habitually take the underground you don't look at the clock because you're used to the regular intervals between trains. However, if you're taking the train for a longer journey, it's important for your own personal orientation to know when the train leaves and when it will arrive since the frequency in this situation is not simply a given. The input interval of the queue replenishment meetings and the output interval of the release planning meetings needn't necessarily be identical (Fig. 6.2). If a Kanban team has established a fortnightly release frequency, for example, then the release planning meeting should take place every fortnight. The queue replenishment meeting can nevertheless take place weekly or even daily if the team's throughput so demands.

The release planning meeting is to be attended by everyone required for the release or interested in the present release, for example:

- Configuration managers and network and operations experts
- Developers, testers, and business analysts
- Direct superiors and management

By the way, releases, even those without Kanban, work best with checklists. Initially, all necessary points are gathered together in a checklist that must be completed in order to guarantee successful delivery. These points are then adapted for the present release and the current conditions established in the release planning meeting.

6.3.1 What Is a Good Delivery Cadence?

We must ask ourselves another question in order to be able to answer this question: When is a delivery economically expedient? Releases don't come for free. Depending on the type and complexity of a project, the costs of delivery can make up a very considerable part of the total costs. In most cases, two types of delivery costs must be considered:

1. **Coordination costs**: comprise all the activities required for the coordination of a delivery including the release planning meetings themselves.
2. **Transaction costs**: before software can be presented to a customer, tests must be carried out, database schemes migrated, servers configured, and backups created.

In order to ascertain delivery efficiency, we calculate the proportion of the total costs the delivery costs constitute. Suppose the costs for a fortnightly delivery cadence in a project with total costs of €100,000 are €80,000. The delivery efficiency is 20%, that is, 80% of the costs are delivery costs. Is this good or bad from an economic perspective? Simply answering "yes" or "no" to this question is not possible because perhaps this is a project for which delivery is very demanding or important. Each organization must thus find its own answer to the question of the most desirable efficiency level. We however have two possible alternatives should it be absolutely necessary to raise the delivery efficiency:

1. **We deliver later**. In this case, the delivery costs as a proportion of the total costs most likely remain the same. If we increase the delivery cadence in our example to a monthly one, the delivery costs as a percentage decrease and the delivery efficiency increases proportionately. The problem with this is that the delivery efficiency would theoretically be at a maximum if we were not to deliver at all. And economically, that's probably not such a great idea.
2. **We reduce the costs of delivery**. Many online platforms can afford to have multiple releases per day because their delivery costs, put bluntly, consist only of a mouse click. Reaching this point however takes a certain amount of work. Where is the potential for savings in the case of other organizations? This second option again makes direct demands on the ability of the system to improve. What we can do to reduce the costs of delivery depends upon the individual situation.

Kanban is thus completely in favor of a regular delivery cadence. However, it doesn't prescribe which of the two paths of increased efficiency we should tread and what the optimal delivery cadence is. This is because, due to the huge differences in demands, it is neither possible nor sensible to make such statements.

Figure 6.2 illustrates the relationship between the queue replenishment meeting (input), the daily stand-up meeting (throughput), and the release planning meeting (output). Input frequency and output frequency do not necessarily need to be in sync. The most important factor in finding the appropriate cadence should always be economic expediency.

This is one of the reasons Kanban doesn't feature iterations in predefined timeboxes (which of course doesn't mean that you can't work with timeboxes!). Work items cannot always be distributed in such a way that they fit into timeboxes. One counterproductive reaction to this is that the work items are simply made to fit so that they can be completed within a timebox. Kanban separates the prioritization of work items in the queue replenishment meeting from development and delivery. The delivery cadences don't need to be fixed to the prioritization cadences if they don't correspond to the way the team works or are not economically expedient.

6.4 TEAM RETROSPECTIVES

Retrospectives are always concerned with reviewing work, particularly collaborative work, taking place within a specific time frame and drawing conclusions on what can be improved. They therefore take a very targeted look at the way the team works. Retrospectives are usually held with a defined frequency. However, as we have now learned, improvement should happen *continuously* in Kanban. As soon as a problem appears, one thinks about how it can be solved and how the work process can be optimized. From this Kaizen perspective, retrospectives are manually created departures from everyday life through which an artificial distance can be created between the appearance of a problem and its educational value. However, realistically, it is not the case that everyone is in a position to immediately implement Kaizen thinking in their daily work. Before there can be a culture of continuous improvement, you first need a culture of improvement. This is one of the reasons Kanban also places emphasis on evolutionary change when it comes to retrospectives. Weekly retrospectives are very sensible when a team begins working with Kanban. In these meetings, the team, management, and other involved groups of people collect suggestions for improvements and sort them in order of urgency. Many teams find it useful to install an improvements board (or improvements backlog) in order to implement improvements one after the other. We have also seen teams use a separate class of service for improvements.

However, retrospectives are also sensible if a Kanban team already very successfully employs Kaizen thinking. While focusing on the current local optimization in a section of the value chain, you mustn't lose sight of the fact that everything must happen in terms of the big picture—the entire value chain of your organization. Seen from this viewpoint, retrospectives are very helpful because a team can take a step back and establish whether the need for all-encompassing improvements at many points in an organization have become visible due to local improvements. In lean production, Kaizen refers to evolutionary, incremental improvements. The concept "Kaikaku" in contrast stands for large, revolutionary, and radical improvements. If a Kanban team has already integrated Kaizen thinking into its everyday processes, retrospectives can become Kaikaku events where the recognition of local improvements can be aggregated into improvements at an all-encompassing, systemic level.

6.5 OPERATIONS REVIEWS

The big picture is also at the heart of operations reviews. Every Kanban team gains valuable experience during its work—experience that leads to self-improvement but that ultimately should also improve the entire organization. With operations reviews, we try to overcome the fragmentation from which organizations "suffer," described by Eliyahu M. Goldratt. In operations reviews, all the Kanban teams in an organization come together in order to share their insights with each other and clarify problems and relationships that exceed the boundaries of the individual teams. The management and stakeholders are expressly welcome to attend in order to get an overview of the progress being made and a better feeling for where their particular skills could be of service. Ideally, operations reviews should take place on a monthly basis and last around 2 h. Strict time management and good moderation are mandatory. During this meeting, each team presents the measurements with which they monitor their workflow and the conclusions that can be drawn from them. The retrospective is thus an objective, data-driven operation for the efficient performance of the organization.

WHAT YOU CAN TAKE AWAY FROM THIS CHAPTER

One of the most important success factors for Kanban is communication: within the team, between teams, and with stakeholders. This communication happens through the visualization of the work process on the board. Having meetings on a regular basis is expedient but not an absolute necessity.

In the daily stand-up meetings, the team discusses and coordinates the current work items each morning. The queue replenishment meeting serves to fill the input queue regularly with work items. Here, the participating stakeholders agree on the order of the work items. The release planning meeting is attended by all those necessary for the release as well as those simply interested in it. The frequency of these planning meetings depends on the delivery cadences established in the process of working with Kanban. Regular delivery cadences strengthen customers' trust in the team.

Team retrospectives are sensible, especially at the beginning of your work with Kanban. The team reviews the work within a certain period of time and analyzes where improvements are needed. In the operations reviews, all the Kanban teams in an organization meet up in order to share their insights and clarify problems and relationships that exceed the boundaries of the individual teams. The retrospective is thus an objective, data-driven operation for the efficient performance of the organization.

7

METRICS AND IMPROVEMENTS

Imagine you are an admiral of the fleet of the British Empire in the eighteenth century. You have just defeated the French at the siege of Toulon, and your men are celebrating wildly on the decks of the 21 ships you command. The formation passes Gibraltar and sets a course for home. As an old sea dog, you know how treacherous the English seas can be. You may have beaten the French, but as the fog slowly slides out over the water, you can do nothing but put your faith in God and prepare for the slow journey. So you summon all of your navigators and have them determine the position of the fleet as precisely as possible. And then it's there. The fog on the evening of October 22, 1707: the god-awful screech of wood on rock, screams, the splashing of the water as hundreds of arms desperately wave around seeking buoyancy. And then it's gone again. The flagship, the Association, has disappeared within a few minutes. Shortly afterward, the Eagle and the Romney sink too. All told four ships go under and with them 1450 men before the fleet can come to a stop. "The longitude! We miscalculated the longitude," you think to yourself as you are washed up at Porth Hellick Cove on St. Mary's Island, more dead than alive. The westward-lying Isles of Scilly were upon you sooner than the navigators had calculated.

You were Sir Cloudesley Shovell just now. Along with his men, the decorated British admiral of the fleet fell victim to one of the greatest problems of sea travel right up till the end of the eighteenth century: longitude. Seafarers since ancient times have been able to determine their position in terms of latitude. But longitude, necessary for a truly precise description of one's position, had astronomers pulling

Kanban Change Leadership: Creating a Culture of Continuous Improvement,
First Edition. Klaus Leopold and Siegfried Kaltenecker.

out their hair. The stars alone offer no reliable fixed point to express a change in position. Many sailors lost their lives before it became clear that the longitude should be understood as a metric of time. The Earth completes one rotation around its axis every 24 h, equivalent to 360° of longitude segmenting the Earth's surface. The Earth therefore rotates 15° per hour. In order to calculate your current longitude, you need the time at your home port as a reference and a clock on your ship that shows the local time—this clock can be reset at midday based simply on the position of the sun. Seven years after Admiral Shovell's terrible end, the British government passed the "Longitude Act" and announced a prize for a solution to the longitude problem using precise time measurements, since the first ship's clocks by Christiaan Huygens were still too unreliable and unstable for use on board ships. Sedate pendulum clocks slipped out of sync due to the ship's swaying, and the sea air also affected them adversely. It was only in 1759 that the woodworker John Harrison, an amateur watchmaker and precision fanatic, created a clock that was accurate enough but also, crucially, easy enough to transport and hold that it solved the longitude problem. After centuries of trial and error, a metric and appropriate instrument had finally been found that could provide change altogether [1].

7.1 METRICS IN KANBAN

Metrics don't only give us our orientation on the high seas. They also help us in our daily life to determine our current location and know how far we have moved during a specific time period. Above all, they can show us whether we have improved or got worse at a particular task. They free us from the danger of only ever being able to measure ourselves by gut feeling. In Kanban, metrics are applied in order to create change and improvement. Just as in the nautical world the position of the ship as a work system is ascertained in order to give the correct instructions to officers and seamen, so too in Kanban do we measure the performance of the system, more specifically of our section of the value chain. We do not focus on the performance of individual employees: if the performance of employees is used as a measure, the individual must always be understood as being under the influence of the system in which they are working because the system creates the prerequisites that determine whether the individual is able to complete the work. If we want to improve a system, we must first of all establish whether this system is built in such a way that it makes efficient work possible. If we improve the system, this will have an impact on the work performance of the individual employees. When measuring the performance of the system, we thus primarily want to see:

- Whether the system works in the way we expect
- Where we can improve things
- Whether our improvement measures are productive

However, measuring the system shouldn't turn into a high-performance exercise. It doesn't make sense to measure everything possible simply because we can.

Conversely, it makes just as little sense to make every metric used excessively precise. We should measure that which is useful and has value. Organizational theorist Russell Ackoff very succinctly stated: "it is better to use imprecise measures of what is wanted, rather than precise measures of what is not" [2]. The number of possible measurements is infinite. Defining the ones to use ultimately depends upon the goals being pursued. In many cases, one goal will be to provide reliable information and commitments to internal and external customers. The measurements show how reliable our information actually is and simultaneously the variables that have an influence on them.

So where should we start? Recommended basic measurements for this purpose are those that allow assertions about the continuity of the workflow, that is, the lead time, throughput, and flow efficiency. These provide information at the outset of your Kanban initiative that is drawn upon in targeted optimization as well as in reductions of variability. Variability, deviating from the standard flow, has an impact on the computability of the work and is caused, for example, by WiP limits that are too high, varying complexities of tasks, a large number of blockers, or the presence of too many bugs. A further issue Kanban teams tend to focus on for improvement is the reduction of waste or, otherwise put, actions that don't actually generate any value for the customer. Exceptions are actions that have a negative effect on the performance of the team if not done.

MEASUREMENT AS A COMPETITIVE SPORT

Riding a tram in Zurich, we were able to—ok, obliged to—observe something interesting: two elderly gentlemen in the row in front of us were having a heated discussion about what could be counted as turnover and what couldn't be. It turned out that their organization's targets for the coming year were up for discussion at the time. Apparently, the two gentlemen had been awarded very ambitious turnover targets and were now trying to identify what one could convince the bosses counted as turnover. Among the ideas, there were such suggestions as "any presence in one of the branches of the organization is counted as turnover," "assistance for internal customers," etc. They were incredibly creative as they tried to optimize the relevant measurement, the turnover target.

A Bug's Life

We repeatedly experience the same thing with software teams. Let's consider that customers are dissatisfied with the quality of the delivered software. The team leader therefore commits the developers to reduce the number of bugs. The team leader uses metrics to decide whether the team is heading in the right direction or not. "Number of bugs in the production system," let us say, is the measurement of choice to be optimized. Sounds reasonable! The plan says that the developers should focus their undivided attention on quality and customer satisfaction will skyrocket.

With a battle cry of bug minimization, everyone gets to work. Pretty soon, bizarre discussions between developers and those responsible for the product can be heard: "That's not a bug! You've obviously described the feature wrong." Why is this important? Well, of course, the developers are supposed to be reducing the number of bugs in the production system. It is therefore very important to know what constitutes a bug and what doesn't. And every successful discussion seems to bring them closer to their goal of fewer bugs in the production system.

But it's not the measurement that needs to be optimized: it's the product!

And wait. What was the initial reason for measurements to be introduced? The customers were supposed to receive a high-quality product. Will the quality of the product improve with the discussion of what is a bug and what isn't? Obviously not. The result:

The measurements will be optimized rather than the quality of the product.

Moreover, a fantastic finger-pointing culture would evolve. "It's your fault, not mine!" The two gentlemen in the tram tried just that. They were concerned only with developing elaborate chess moves so that the measurement "target turnover" could be reached at the end of the year.

It is always the case that measurements are a genuinely good thing. But only if the goal of the measurement is firstly clear beforehand and secondly actually being pursued.

7.2 CUMULATIVE FLOW DIAGRAM

One of the most powerful metrics for workflows is the cumulative flow diagram (CFD). As the name implies, it shows how well a system flows (Fig. 7.1). How do we create this diagram? The x-axis is time. Using the stipulated unit of time (e.g., days), the number of tickets on the board is entered cumulatively (y-axis), subdivided into the process steps. Important information should be entered next to the data to aid observations and the identification of relationships once the graph is drawn. A system that flows well is illustrated by lines having a constant positive gradient and channels that represent the individual lines maintaining a constant width. Kinks in the line indicate bottlenecks at those times. Using this temporal information alongside collected data about events like change of WIP limits, blockers, holidays, etc., we can more precisely research the problem a system had on a particular day or for a particular period of time.

Vertically, we are able to read off the work in progress (WiP) for each process step, that is, how many tasks per process step are currently being processed, in the input queue (in this diagram given as "to do"), or already completed. Horizontally, along the x-axis, we can extract the average cycle time for the tasks.

Let us examine what information can be extracted from the CFD in Figure 7.2. Between weeks 2 and 3, we can see that work in this system was done without WiP

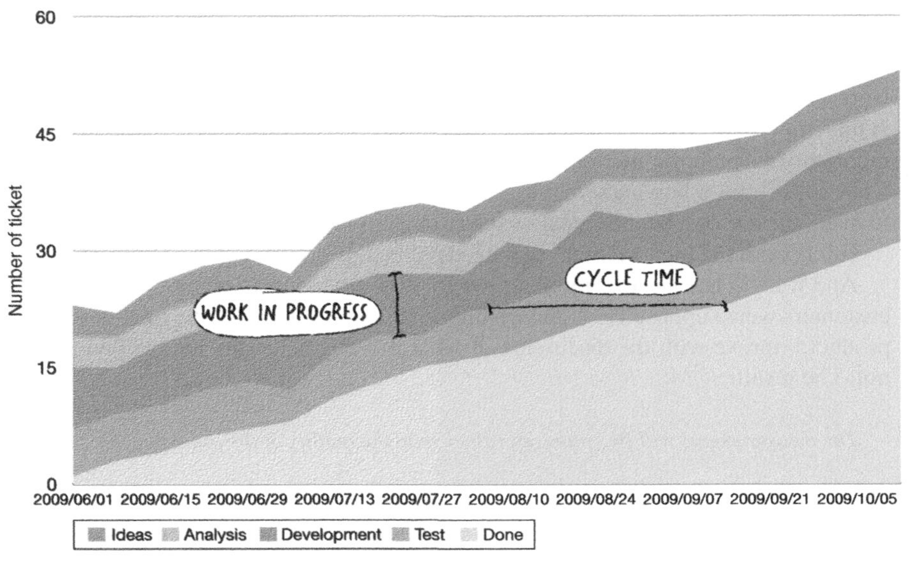

FIGURE 7.1 Cumulative flow diagram.

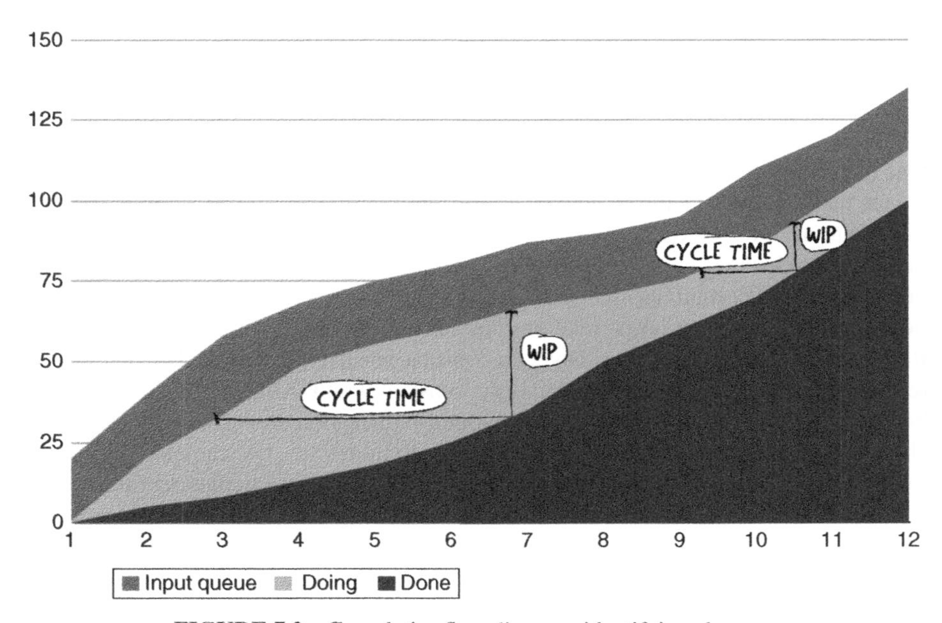

FIGURE 7.2 Cumulative flow diagram: identifying changes.

limits: the WiP became correspondingly larger, and the curve for the completed tasks grew with a very small gradient, that is, new tasks were only completed very slowly. In week 5, the decision was obviously taken to apply WiP limits: from this point forward, the curves for current and complete tasks have steeper gradients than before. This means that the WiP limit has a positive effect on the cycle time of the work. Without WiP limits, the team would have started working on more and more tasks while completing ever fewer. The relationship between WiP and cycle time in weeks 4 and 10 is remarkable: on average, the cycle time in week 4 is 4 weeks and in week 10 is 1 week. From this, we can say:

The higher the WiP limit, the longer the cycle times.

7.3 MEASURING THE LEAD TIME

On many occasions, we would like to be able to tell the client very precisely how long it would take to process a standard ticket, for example, or the probability of a ticket being completed within a specific time frame. We must track the lead times of the tickets in the individual class of service and task types in order to be capable of doing this.

The simplest method is to enter on a diagram the time at which a ticket is completed for each service class individually or for all classes of service together (Fig. 7.3). The *y*-axis shows the lead time and the *x*-axis shows the completed tickets. It is also advisable to add a trend line for the measurement. A very easy option is to calculate a moving average, where the average is calculated for every three points and entered on the graph as a connected line. In this way, after a little time has passed, we can say whether the lead times have remained constant or become longer or shorter. The reason for particular changes in the lead time can be identified if the diagram is examined with reference to other information

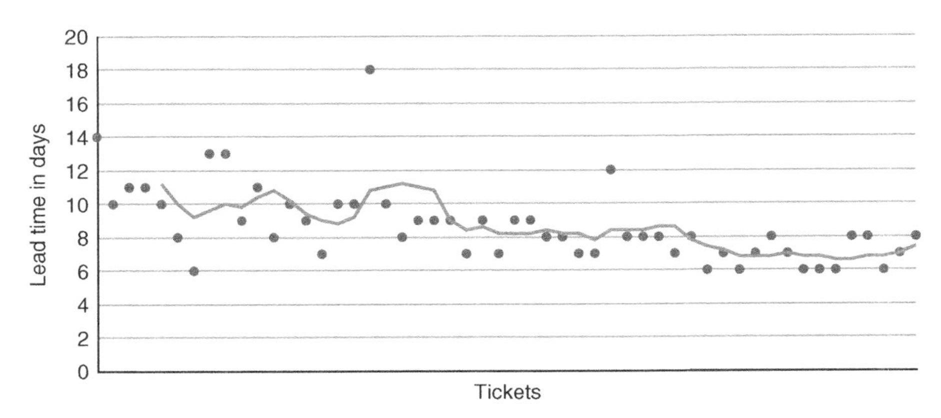

FIGURE 7.3 Measurement of the lead time including a trend line.

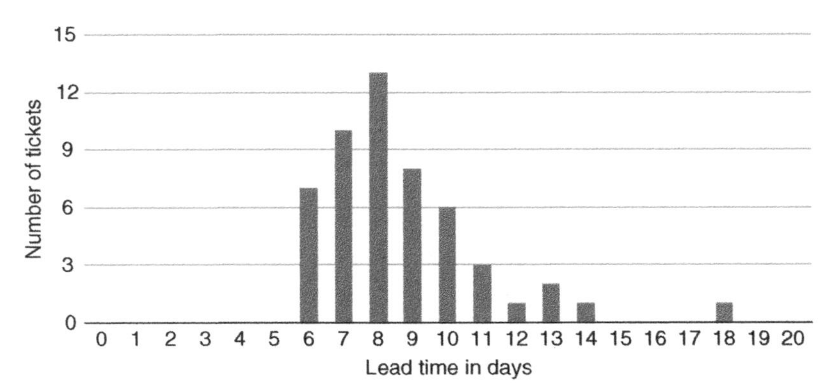

FIGURE 7.4 Histogram of the lead time.

collected previously concerning our process. An image of the impact of decisions, bottlenecks, blockers, etc., thus evolves. As with the CFD, with this graph, we will see that the lead time becomes longer if we increase the WiP limits and shorter if we decrease them.

Frequency analyses using histogram charts of lead times are very useful for the calculation of the service-level agreements. For purposes of orientation, we enter the lead time in days on the x-axis and the number of tasks available on the y-axis. In Figure 7.4, for example, we can see that we needed a lead time of 8 days for 13 tasks. We can also see an outlier ticket that took 18 days to process.

Altogether, we have measured 52 tickets in total, of which five had a lead time of more than 11 days. Thus, we can say that tickets of the class of service represented are 90% of the time completed within 11 days. In order to be able to provide precise information, the distribution of the completed tickets should lie within a small range and have few or no outliers. The broader the distribution, the higher the variability—deviation from the norm—the system exhibits. Suppose the bar at the lead time of 18 days is larger. If we have collected the tickets from the board, we can study them more closely and look for the cause of this high lead time. There could be a common cause, and as a result of this, a new task type is added to our board.

7.3.1 Throughput

If we measure the lead time, the throughput can be simultaneously measured "for free." The throughput describes the number of tickets delivered within a certain period of time. In contrast to the lead time, of course, the throughput should increase. Here, we again have a direct link: the lower the lead time, the higher the throughput, and vice versa. It is perfectly possible, in the context of measuring the throughput, to measure how meeting service-level agreements has developed. In Figure 7.5, the throughput in January was 14 tickets: 10 of these were delivered on time, and in the case of four, the service-level agreements were not met. If the main goal is still customer satisfaction, we should pursue the due date performance exclusively.

FLOW EFFICIENCY

Flow efficiency is a temporal measurement that is already rather advanced. As a measurement, it is highly informative but too energy consuming and not particularly suitable for daily manual tracking. In this case, support from electronic tools is useful since the flow efficiency is a result of the measurement of three values:

1. The **processing time** describes the time in which the tasks are *really* worked on: the ticket was not blocked and not in any other area, waiting to get taken up, during this time.
2. The **waiting time** is the time a task spends in a "done" queue, waiting to be handed over.
3. The **blocked time** is the time a ticket was on the board but couldn't be processed because it was blocked for some reason.

The flow efficiency is therefore the percentage of time during which work is done on a ticket on the board: a comparison between the entire lead time and the processing time (processing time divided by lead time, multiplied by 100). When you begin to make these measurements, the results are sometimes initially shocking, especially in systems without WiP limits. You suddenly notice that only 5% of the time is spent actually working on a ticket, while the ticket spends the rest of the time being blocked or in queues. We can see that this metric is ultimately suited to visually representing the change that happens over a long period of time. However, getting worked up with shocking flow efficiencies on a weekly basis would be counterproductive and wouldn't say very much about improvements or advancements that require an amount of time to have any effect.

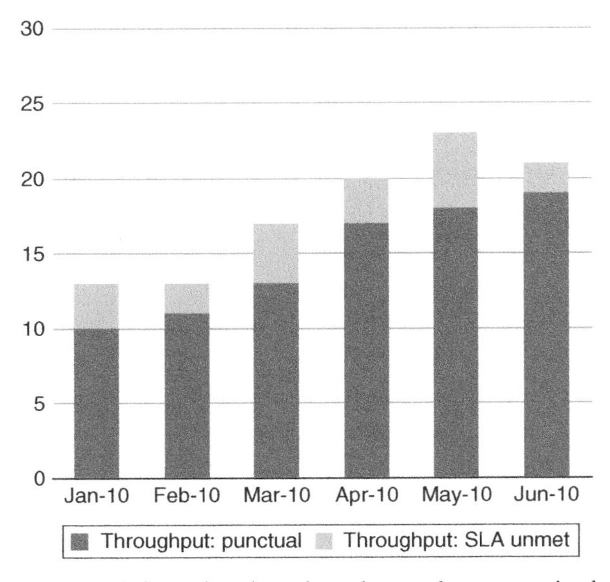

FIGURE 7.5 Combined information about throughput and unmet service-level agreements.

7.4 REWORK AND BLOCKERS

Measuring the failure load establishes the number of tasks that are in fact already completed but, due to poor quality, need to be reprocessed.

This is an essential measurement: everything that is not done properly lands again at some point in the input queue, possibly as a bundle of expedite tickets. If speed is valued above quality, a team runs the risk of disrupting its workflow either in the short or long term because it must once more use every trick in the book for these exceptions in order to neutralize the additional backlog. The problem is that there is normally a large chunk of time between a bug being delivered and discovered. If an error returns to the system, you first need some time at the beginning to reacquaint yourself with the work and then time to find the bug. The number of rework tickets should decrease with time.

You will often hear teams say, "Ah, let's not test so precisely right now. Let's go straight to release." But ultimately this has the same effect as giving blood to lose weight. You do indeed lose weight initially, but after a short time, you're still staring at the same number of kilos on the scales.

7.4.1 Blockers

As opposed to the failure load, blockers appear while tasks are still being processed. In order to make them visible, they are marked on the board with a red sticker attached to the task with the corresponding information on it, for example. As improvements are made and the blockers eliminated, it is obviously wise not to throw away these stickers. Collecting, clustering, and analyzing these blocker tickets to determine the reasons for their presence will produce some new approaches for the improvement of the lead times, because blockers obviously increase the lead time. Through this process of collection, the number of blockers can also be represented along time (Fig. 7.6). And if the blockers are analyzed

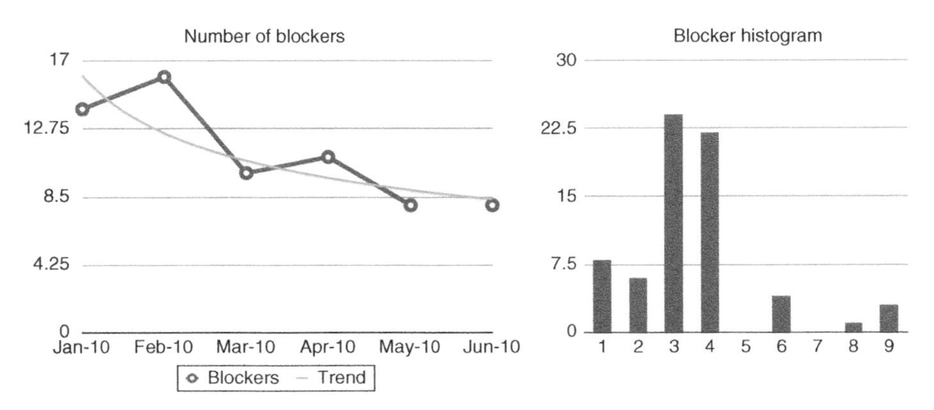

FIGURE 7.6 Measurement of blockers.

sufficiently, the causes identified, and the right treatment found, then the curve should describe a positive trend.

Just as in the case of lead times, a histogram is also a fascinating instrument for describing blockers. We can use it to establish how many tickets were blocked and for how long. From the representation in the diagram, we can draw conclusions about how agile an organization already is: Are blockers eliminated quickly, or does their removal require a couple of weeks? In which departments and areas do the blockers appear? Do the blockers perhaps appear at regular intervals?

7.5 IMPROVEMENTS

Measurements show us where the system is "getting caught," how it has developed, and where we need to change things in order to reach the stipulated goals. Measurements bring us onto the right track: they are the first step. But in order to be successful, we must of course make actual improvements. There are a huge number of approaches for this, one of which we have already seen in the theory of constraints in Chapter 4. However, we must ask the following question before applying any special methods of improvement: Why is the situation the way it is, anyway? Because, for example, the theory of constraints is too drastic an approach or perhaps even the wrong theory altogether if we have temporary bottlenecks in our system that only appear in the holiday season. *The basis for all improvement approaches is thus a deep understanding of the problems.*

Root cause analysis presents causes and effects in direct relation to one another. We could be very academic and plot cause-and-effect graphs. However, there are far simpler and therefore frequently more effective methods of implementation. If the same undesired situations (symptoms) are always observed in a system, the simplest and most expedient form of root cause analysis is to ask oneself why a few times. You begin at the level of the problem apparent to everyone and work step-by-step backward until you reach the actual cause.

WHY?

Let us examine a supposed problem that is in actual fact a possible solution. In the context of a Kanban implementation, we held interviews with the stakeholders in order to identify the problems the organization was experiencing. In one of the interviews, someone mentioned the problem that no unified calendar system was in use.

Statement: "We don't use a unified calendar system."
Question: "Why do you need a unified calendar?"
Answer: "Because each person enters their appointments in their own calendar."
Question: "Why is that a problem?"
Answer: "Because we don't know which appointments our developers have."
Question: "Why do you need to know which appointments your developers have?"

Answer: "So that we can see when they have free capacity."
Question: "Why do you need to see when they have free capacity?"
Answer: "Because we want to know when the developers are working."
Question: "Why do you want to know when the developers are working?"
Answer: "Because we damn well want to know when our stuff is finally going to be ready."

In actual fact, there are far more effective solutions for this problem than a unified calendar system. Beware of logical leaps—they appear all too often in practice.

Only once we've identified the causes of a problem and properly understood it does it make sense to think about the possible ways of solving it. In most cases, it is a question of a coming together of several problems; thus, the question that keeps arising is whether a possible solution solves the "right" problem—does it really produce a lasting change, or does it only put the symptoms temporarily on hold? Albert Einstein once said, "If I had an hour to solve a problem and my life depended on the solution, I would spend the first 55 min determining the proper question to ask, for once I know the proper question, I could solve the problem in less than 5 min."

7.5.1 Theory of Constraints

Remember, the theory of constraints is based on the systems theory observation that the throughput of a system is exclusively determined by one limiting factor (Chapter 4). We can only improve the throughput if the whole system starting with this limiting factor is comprehensively optimized. To this end, we have mentioned Eliyahu M. Goldratt's five focus steps, which form the core of our treatment of bottlenecks. Only one of these five steps is ever used in many organizations—the attempt to widen the bottleneck by piling more resources into it. In Goldratt's scheme, however, this is the fourth of the five steps and should only be done once the first three steps are complete.

Fact is projects do not get quicker if more resources are invested on a short-term basis. The additional resources must firstly be found (e.g., new employees). Then the new resources need an amount of time to become established, for example, training new colleagues. Such actions are more likely to *increase* the lead time further rather than *decrease* it. Or, as stated by Frederick P. Brooks, adding resources to a late project makes it later [3]. What potential improvement can the theory of constraints bring in the context of Kanban if the five steps are properly followed?

1. **Identification of the bottleneck**. Kanban makes this step very easy for us because we can quickly identify where things are clogging up the system using our visualization. We must however remember the root cause analysis once more. Do we have a real permanent bottleneck on the board in front of us?

Only once this question has been answered should a bottleneck actually be treated as such.

2. **Taking advantage of the bottleneck**. The optimal use of the bottleneck is through a combination of relieving and applying load:
   a. **A bottleneck must be relieved of all tasks that don't absolutely have to be done in it**. This means, for example, the reduction of administrative tasks that don't have any direct correlation to the job's results. Blockers should be solved by employees from outside the bottleneck so that the employees in the bottleneck can concentrate fully on their actual work. Apart from this, risk profiles of work in the bottleneck can be established so that some of them can be taken over by employees outside the bottleneck.
   b. **The bottleneck must not run idle**. Consistency in the distribution of tasks is necessary for the throughput of a bottleneck to remain stable. Idle periods in the bottleneck are just as bad as permanent overburdening. In order to avoid idle periods, we build a buffer in front of the bottleneck that evens out variability in the upstream stages—such as if resources are not always available. This buffer should be as small as possible but large enough to establish a smooth and consistent workflow.
3. **Working with what is already in place**. We ensure that there is only ever as much work in the entire system as the bottleneck can process, that is, we set the WiP limits accordingly.
4. **Removing the bottleneck**. If due to the previous three steps we can assume that things in the bottleneck are working optimally and that we have ensured a constant workflow with our WiP limits, then we can take another look at the bottleneck. How does the bottleneck look after these initial measures? What is now necessary to do to remove the bottleneck permanently if it is still present? Only at this stage do we do what most organizations immediately do: invest resources.
5. **Starting again at step 1**: In the summer of 2011, to the dismay of many holidaymakers, the theory of constraints manifested itself in a very clear example. The Austrian Tauern Road Tunnel has always been a bottleneck on the connection between Northern and Southern Europe. Record queues were observed every summer until the decision was finally taken—not least because of the catastrophic fire in 1999—to build a second tunnel. This second tunnel was opened on June 30, 2011. Everyone was happy because the traffic rolled through nice and quickly…right down to the one-tunnel Karawanks Tunnel a few hours later, where queues up to 60 km long built up. Such a phenomenon doesn't only appear in traffic bottlenecks: it's also something we have to consider in organizations. If we have eliminated a bottleneck at one point, another could appear at another location. In a figurative sense, we research the bottlenecks in the organization and eliminate more and more of them until we reach the point when the bottleneck finally no longer appears in the organization but is in fact shifted over to the market itself. In an ideal world, we would supply quicker than the market can demand.

7.5.2 Reducing Waste

In Kanban, we don't simply want to deliver more quickly: we also want to develop our work economically. The costs of delay that constitute the backdrop for classes of service, for example, are one such approach. In the release planning meetings (Chapter 6), we also try to find an economically sensible delivery cadence and save time with checklists. Key to good economic thinking is to know what is *not* economical, that is, what can be described as "waste." In software development, this means all activities and/or expenses that don't bring any additional value for the customer. Principally, this would be large piles of partially completed projects, task switching, or long lists of tasks—much of this can either be eliminated or optimized using the Kanban approach. But features a customer has neither ordered nor needs or which developers often only include because they're thrilled with their own abilities can also be considered waste. There are then two further categories of waste we can tackle:

1. **Transaction costs**: These include the preparatory and follow-up projects that accrue around a project, that is, setup activities, project planning, resource planning, risk planning, budgeting, delivery, user training, retrospectives, reviews, etc.
2. **Coordination costs**: These include the sometimes massive expenses that poorly prepared and badly moderated meetings of all types cause. One of the cheapest improvements that can be implemented is to carry out meetings efficiently. Daily stand-up meetings can be reduced to 10 min, for example, and technical questions can be relocated to follow-up meetings in order to not tie up the resources of the uninvolved employees unnecessarily. This frees up time capacities in which work can be done on the customer's project.

The term "waste" is derived from lean production and probably has a somewhat negative connotation for many readers. To reduce waste doesn't mean to mercilessly discard everything that doesn't bring additional value to the customers. It means the optimization of activities that we ourselves must put in place for the implementation of a project, in such a way that the least costs are generated (in terms of both time and money). So we question the value creation of our activities, "Does it produce additional value for the customer if we do more of this?" Three more meetings are of no use to the customer or the organization itself if nothing tangible results from them.

7.5.3 Reducing Variability

By variability, we mean deviations from the standard workflow. The more variability a system exhibits, the harder it is to provide reliable information about lead times. In contrast to the assembly industry with its coordinated machines, knowledge work, due to the human will, will always have a high level of variability. The question is whether this variability should be reduced to a minimum or whether variability perhaps also contains its own potential. We do in fact need a certain degree of variability in our system in order to produce what we termed "slack" in Chapter 4. Due to this variability, there will always be employees who are temporarily "unemployed." This is standby capacity for emergencies, improvers, and innovators who can efficiently resolve the weak points in a system and turn them into strengths, while the others concern themselves with the regular daily work. Ultimately, reducing blurriness is productive only in areas where this blurriness itself causes problems, for example, in cases where we want to make service-level agreements with our customers and have a very high level of variability in the completion of the jobs. What's a customer supposed to do when faced with, "The completion time for the job is somewhere between one and 100 days."? The job could be ready in 10 days, or in 99 days, and we ourselves cannot be more precise. The effect: an angry customer who can't make plans because we can't make reliable commitments.

What are the causes of high variability? Among others:

- Different ticket sizes
- Absence of WiP limits for task types
- Limits on the various classes of service that are too high (e.g., too many express tickets)
- Too much time fixing errors due to bugs
- A clogged-up workflow due to too many blocked tickets

With all the analyses and other instruments that Kanban makes available, we can find out what causes variability in our system and what impact it has. But variability shouldn't be reduced just for the sake of reducing it. If it is disruptive and compromises relationships with customers, it should be restricted to a level that is tolerable. But we must also get used to the idea that variability is an integral trait of knowledge work.

We want to improve performance in the world of work with Kanban. We're not interested in raising programming Jedis able to program ever more in decreasing amounts of time so that more jobs can be taken on or weakness glossed over. Kanban doesn't consider a person a weak point but instead observes the system in which we operate as a network of interactions that sometimes experiences problems. Using its instruments, Kanban turns its users into system thinkers who recognize and understand how different approaches influence each other. Participants begin to think in terms of processes rather than isolated individuals and don't see merely black and white but rather all the shades in between. With time, it will become clearer that everything in a system is connected and has an own role in the overall function.

WHAT YOU CAN TAKE AWAY FROM THIS CHAPTER

Metrics are important in understanding changes and improvements in a team's processes when working with Kanban. In Kanban, it is always the system that is measured and never the performance of individual employees. By measuring the performance of the system, we want to see whether the system works as we expect it, where we can make improvements, and whether our improvements have any effect. Measurements shouldn't be concerned with every last detail but should provide quick and reliable information about what a team wants and needs to know. The measurements applied depend upon the goals a Kanban team pursues.

Recommended basic measurements concern the lead time, throughput, and flow efficiency of a process. Metrics for these include, for example, the CFD, tracking and histograms of lead times, the failure load, and blockers. The results of these measurements are the basis for decisions regarding improvements.

Before implementing specific methods of improvement, we should ask ourselves why the situation is the way it is. Root cause analysis is here an appropriate instrument for the identification of the causes. In terms of improvements, the five focus steps of the theory of constraints help identify (actual) bottlenecks, optimize their use, and broaden them step-by-step. In the case of reduction of waste, we orientate processes around the activities that actually generate additional value for the customer. By reducing variability, we achieve a greater degree of reliability in the information we provide to a team's stakeholders.

PART 2

CHANGE AND LEADERSHIP

8

FORCES OF CHANGE

A young boy walks along a beach littered with thousands of starfish. Every couple of meters, he bends down to pick up a starfish and throws it back into the sea. A man watching the boy walks up to him shaking his head and asks him, "What are you doing?" "I'm saving the starfish," the boy answers. "But that doesn't make any sense," replies the man, bewildered. "What difference does it make to the other thousands of starfish lying here on the beach?" "It makes a difference to this one here," the boy replies, throwing one more starfish into the sea.

"Making a difference" is perhaps the fundamental mantra for change. Similarly, "making lots of small changes" could be Kanban's mantra. It could be that this causes just as much headshaking as the man in the story about the young boy and the starfish. Why concern yourself with individual starfish when a comprehensive rescue project is really what's needed? Because many small changes add up to a big change, we might answer, using the beach story as an example. Because every starfish thrown back is equivalent to a small improvement. And because we, just like the boy, have the power to make these improvements.

Something must obviously change if things are to improve. And, as the story about the starfish shows, change is always a little controversial. On the one hand, it seems to come about of its own accord: a powerful gust of wind, an unexpected sea current, and suddenly the beach is covered with starfish. On the other hand, we invest a huge amount of energy in holding back the natural flow of change in rather artificial ways; we try to predict the weather; we focus our attention on the beach; we build

Kanban Change Leadership: Creating a Culture of Continuous Improvement,
First Edition. Klaus Leopold and Siegfried Kaltenecker.

dams against the breakwater. And we throw washed-up starfish back into the sea. This is how we make a difference, step-by-step.

The German systems thinker Dirk Baecker asserts that "management makes a difference" [1]. According to Baecker, management introduces differentials into the existing procedures. In other words, management provides change. It is nothing new that in today's world management is ultimately change management. The number of Google hits alone shows that change has become a kind of force of nature: about one billion entries for change and more than 500 million entries for change management (as of July 2014).

But why is change management even necessary? What concepts are there for it? What does Kanban's evolutionary change provide? And how do you make the best of the demons that you call up with it? We will tackle these fundamental questions and provide you with the most succinct answers possible over the next six chapters. We will focus on:

- **Forces of change** to give you an overview of the most important factors featuring in professional change management
- **Environment and systems** in order to create the theoretical basis for a suitable understanding of organizations
- **Organizational and personal change** to clarify how corporate change is experienced and processed
- **Emotions in the process of change** to investigate more thoroughly how change is really experienced
- **Corporate culture and politics** in order to understand this way of experiencing things in a broader context
- **Conclusions for Kanban change leadership** in order to create guidelines for a culture of continuous improvement

8.1 TURBULENT TIMES

"For the times they are a-changin'," proclaimed Bob Dylan as early as 1964 as he sang of pressing dangers, the disappearance of the old order and the need to keep your eyes open. "The Times They Are A-Changin'" could easily be used as the title song for the intense change that we've experienced in the last 50 years. The following factors, among others, can be identified as catalysts for this change:

- **New technologies** that enable work at ultrahigh speeds
- **The globalization of work and consumer markets** that radically change supply and demand
- **The individualization of customer desires**, driven by a major new lifestyle orientation in powerful industrial countries and also through increasing standards of living in developing countries
- **The mobility of people and work** that enables totally new networking

- **The internationalization of collaborative work** at economic and political levels
- **The outsourcing of production processes** to so-called cheap-labor countries, which in turn causes the expansion of transport routes and the erosion of social standards
- **Better education** that is more readily available and the evolution of regional knowledge clusters
- **Demographic change**, which forces organizations—in the western hemisphere at least—to concern themselves with the issues of age, background, and culture
- **Aggressive market competition** that puts constant pressure on the individual as well as the organization to perform
- **Rampant insecurity** driven by financial speculation as well as political conflict and environmental disasters

All these developments bring with them new demands. Just like a juggler juggling many balls, highly diverse forces and capabilities must be coordinated simultaneously. Of course, specific forces are particularly relevant to each organization: some have a massive impact on the current business situation, while others remain in the background.

Figure 8.1 illustrates the network of forces of a financial services provider whose current drivers of change are represented by varying their size and distance from each

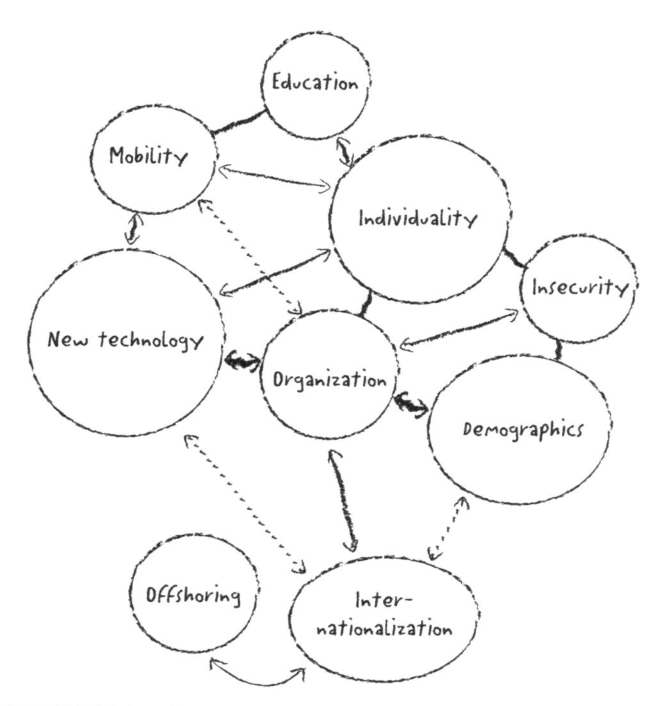

FIGURE 8.1 Current challenges for a financial services provider.

other. New technologies are significant for this organization, mostly due to the virtualization of many service branches. Individualization is reflected in the custom-made products demanded today, and demographic change points to a future in which age and migration will play a significantly larger role. The increasing insecurity of people and the competitiveness of the market lurk as it were in the background, while internationalization and offshoring currently have no strategic relevance.

The most important environmental factors demand corporate measures such as the development of a new CRM system, the equipment of all traveling employees with iPads, or diversity training programs in order to better understand the often underestimated demographic change. All involved should of course bear in mind that this diagram of the network of forces is only a temporary record—the half-life of this image is short. Due to our tense economic climate, new forces can come into play at any time, demands change in a flash, competing firms pop up in highly profitable niche markets, and customers find what they need on the Internet themselves. On top of all this, there are incalculable knock-on effects not only between the organization and its most relevant factors of change but also between the factors themselves. As the financial services provider perhaps somewhat intensely experienced during the last few years, increases in mobility influence technical demands as well as the nature of consultancy itself. Individualization changes supply as well as demand. Older customers and customers with migrant backgrounds represent new demands and challenges.

For a long time now, organizations have no longer been able to choose whether they want to participate in this power game or not. Change has become mandatory.

Trying to hold onto the status quo is like trying to keep the leaves on trees in autumn. As the following overview shows, not changing in today's world exacerbates a multitude of risks.

DANGERS OF NOT CHANGING

Economic:	If you don't change, you will go out of business or your share of the market will at least decrease.
Political:	If you don't change, you will lose out to a more powerful group.
Technological:	If you don't change, you lose your connection to future-defining standards.
Legal:	If you don't change, you must be afraid of restrictions.
Moral:	If you don't change, you are considered socially irresponsible.
Personal:	If you don't change, you lose recognition, influence, and maybe even your job [2].

For an organization to be successful, it must overcome these dangers and use the opportunities every change brings along. In other words, the organization must keep up with, or ideally be ever so slightly ahead of, the current market demands. It's a shame then that this market behaves unpredictably. That which is "top" today can be a "flop" tomorrow; yesterday's success factor can become a burden overnight. "Business agility" will become the mantra for the successful running of an organization in the twenty-first century. Improvement and innovation have long since become mandatory for any organizational unit. Available opportunities should be used, new possibilities discovered, and competitive edges honed.

Although this mantra is repeated by almost every organization, statistics show something else entirely. According to current studies, only one in every five employees is fully engaged, 75% of all employees lack motivation and passion, and only 15% of all teams are able to realize their full potential. Furthermore, as management pioneer Peter F. Drucker demonstrates, many organizations still look at change as they would death or taxes: unwanted but unavoidable and to be put off as long as possible [3]. However, even the change projects taken on board seldom achieve the intended goals. There are no comprehensive figures but various sample surveys point relatively clearly to a proportion of between 60 and 80% of projects ending in failure.

Following this, we shouldn't forget that, with the current environment, many change efforts fall victim to turbulence. Just like the future, the change is no longer what it once was. The number and variety of change initiatives have increased tremendously, while current rationalization and innovation projects, outsourcing and merger measures, reduction of employee numbers, and targeted acquisition of expertise have become in many organizations the norm. As Table 8.1 shows, the "Dance of Change" [4] in today's world must consider a huge variety of conflicting priorities.

TABLE 8.1 Conflicting Priorities in Change Management

Long-term planning	
Changes require time	Short-term results
Strategic parameters	Pressure to make changes quickly
Focusing on clear goals	Openness to unplanned processes
Parameters of project management	Focusing on flexibility
Step-by-step improvement	Creative room for action
Customer or market oriented	Radical innovation
Managing risks	Employee oriented
Precise problem analysis	Capitalizing on new opportunities
	Quick fixes

8.2 TURBULENT CHANGE

Everything considered, implementation pressure has grown tremendously: plans for change must be implemented quickly, or else, they run the risk of becoming obsolete before they've been implemented across the whole system. This makes the time window for the necessary mobilization just as tight as the one for shared reflection, not to mention the continuous adaptation of the change process from what was originally conceived. Thus, repeatedly revisiting things becomes a characteristic organizational experience of the twenty-first century, according to German economist Günther Ortmann [5]. Like some never-ending somersault, market developments spin around and around their own axes in accordance with organizations' attempts to keep up with these developments. In this era of hypercompetitiveness, there's scarcely any time left to come to rest, demonstrate value, or fine-tune things. Before change projects can properly deliver return on investment, the next wave of change rolls over them.

The speed of change procedures doesn't just pull the rug from under the protagonists' feet; it also prevents attainment of the originally stipulated purpose.

The new developments can barely be internalized and integrated in the day-to-day work. Thus, a situation quickly evolves where, according to one manager we know, changes "pop back up and have to be chewed on again." It's no wonder. In the area of responsibility of this management friend of ours, in his role as CTO in a telecommunications organization, seven different change initiatives are running simultaneously: two initiatives for process improvement, two training and coaching projects for the strengthening of individual leadership competence, one workshop series for cultural integration following a merger, one team-building program, and last but not least the introduction of Kanban. To make matters worse, the original goals of two of these initiatives have already changed: one of the process improvement initiatives is heading in the direction of collaboration between departments; and, following the acquisition of a rival firm, the training initiative is completely geared toward integration management. The original priorities—the comprehensive improvement of agile approaches and a complementary Agile Leadership Training Program for the management team—have been swept away with the tide. New issues have taken their place, causing many a change agent to break out in a sweat.

"Moving targets" become commonplace. What is of paramount importance at the project start can quickly become a sideline issue as time progresses. Instead, as our friend reliably informs us, additional change requests appear all over the place—requests that should ideally be integrated in the running change procedures: the customer would like new product features; the competition is offering a similar product at a competitive price; and the EU is making legal parameters for online operations stricter, for example.

The experiences of our CTO friend specifically outline what is currently on the to-do list of many organizations. Large organizations must get a grip on their *projectitis* and achieve balance in their myriad change impulses. Various corporate areas recommend the use of very different approaches:

- Continuous improvement with Kanban
- Innovative prototyping
- Six Sigma
- Outsourcing
- Traditional, waterfall-style project management

- Software development with Scrum
- Corporate social responsibility
- Customer value management

Running simultaneous, uncoordinated change projects brings with it a bundle of new interfaces as well as the risk of coordination conflicts. Furthermore, the lack of coordination between the change initiatives creates a type of wave pattern that makes the surface of the organization look like it's moving dynamically, while underneath nothing's really happening. Instead of changing settings, these change waves are routinely ridden out.

> *You don't have to be a fortune-teller to recognize an overwhelming change fatigue in many organizations. Rather than commitment, many projects are more met with an attitude of "not again!"*

Weariness and a high level of not taking things seriously are the rule of the day. These attitudes are also encouraged by the fact that many organizations react to crises in their environment ineffectively. Solutions are discussed before there is even agreement about the problem; the current challenges are not diagnosed together but rather predetermined; the benefits of the current situation are not sufficiently considered. Alternatively, changes are adopted without consideration of the actual costs. If you think over this small hit parade of failures, it will become quickly apparent that many change projects don't flounder due to bad processes or a lack of tools. It is far more the case that these projects are victims of false, basic assumptions. Need a couple of examples?

- **The problem is crystal clear and all that's needed is a quick fix**. "We don't need a diagnosis," said the CEO of a real estate platform, perfectly sure of himself, when we wanted to explain our tried and tested approach to him during the clarification meeting. "We only need your specialist advice as to how we can raise our development speed and reduce our test errors as quickly as possible. Just tell us what we should do!"
- **Successful change management primarily needs project plans that then just have to be implemented**. "I don't know what's up with them," the production leader said, shaking his head hopelessly while preparations for the next year's examination were underway around him. "Up until now, the shift supervisors were in line with us. But ever since they had to start following the new parameters of quality control, there are new difficulties all the time. We need to talk some reason into them about this exam once and for all!"
- **You just need to bring together the best specialists and experts in order to create successful change**. "Child's play," we can still hear the all too confident voice of an R&D manager saying as we asked him about the degree of difficulty of the desired solution. "I've got my best people on it." That these people hadn't actually got used to working together and didn't have all the necessary skills was only noticed after they had failed to reach the set milestones.

In his book *The Logic of Failure*, Dietrich Dörner expertly analyzed the fact that untested assumptions not only produce shortsightedness but also have concrete consequences [6]. The essence of his analysis is that through leaping into action, overdoing established measures, or denying side effects, change management itself becomes the problem. Unfortunately, many organizations would rather play around with familiar symptoms than research causes as yet unknown to them.

THE LOGIC OF FAILURE

Acting without systems thinking
Tendency to leap into action
Acting without consideration of long-term effects and side effects
Tendency to overdo established measures, particularly under time constraints
Acting without consideration of processes
Tendency to control specific conditions rather than dynamic processes
Acting without precisely seeking the errors
Tendency to professionalism, where you solve what you can rather than what you should
Acting without recognizing your limits
Tendency to expertism, which can tempt you toward an overestimation of your own abilities or the transgression of safety standards
Acting without organizational awareness
Tendency to carry out isolated projects without considering organization-wide needs for change
Acting without critical self-reflection
Tendency to evade uncertainty through denial or delegation of the problem [6, pp. 34–37].

Occasionally, change brings to mind Mark Twain's short story "The Watch" in which a watchmaker who was previously a steamboat engineer says about a defective pocket watch, "She makes too much steam—you want to hang the monkey-wrench on the safety-valve!" A previous watchmaker had reckoned otherwise. "He said the king-bolt was broken." After being successfully repaired, the watch "would run awhile and then stop awhile, and then run awhile again, and so on, using its own discretion about the intervals." Every time the watch "went off, it kicked back like a musket." Mark Twain's pitiable watch owner laments, "I padded my breast pocket for a few days, but finally took the watch to another watchmaker" [7].

Does that sound familiar? Do you also have such "watchmakers" in your organization who in fact have a different profession to the one they're engaged in? Managers who only change what, according to their specific perception, they are able to change? Change agents who only work in their own familiar areas and not in the areas that should be addressed? Who ignore their own blind spots along with the risks that accompany repair measures? If this change logic seems familiar, it will hardly come as a surprise to learn that 60–80% of projects fail and that the average life expectancy of organizations is now less than 20 years.

WHAT YOU CAN TAKE AWAY FROM THIS CHAPTER

In today's world, change is the rule rather than the exception. Professional management of this change should be on the agenda of every successful organization.

A number of external drivers of change should be taken into consideration, which can be defined briefly as globalization, mobility, diversification, and demographic change.

One should also consider that the internal conditions for change projects have changed. Most importantly, the diversity and speed of change initiatives in large organizations have recently exploded.

This explosion is accompanied by an implosion of motivation: weariness, exhaustion, and various forms of protest belong just as much to change work in the twenty-first century as the commitment and passion of many other participants.

Change is often complicated by unsuitable action strategies. Leaping into action, overdoing established measures, and the blanking out of follow-up and side effects are definitely not suitable means to deal with the complexity of the market.

9

ENVIRONMENTS AND SYSTEMS

So it seems that the failure of change is a part of daily business. What can be done against this? How can you best deal with the overwhelming number of change impulses? What will prevent your successful change management from becoming pure fate?

Let us take a little time to answer these questions. Ultimately, well-founded answers are not possible without the appropriate research. "There is nothing more practical than a good theory," said change pioneer Kurt Lewin [1], so let us try to establish our change management on a solid basis. In the systemic worldview, the crucial question concerning successful change management points to your own perception of the organization. What is my understanding of what I want to change? What is an organization, anyway?

When we talk about organizations on a day-to-day basis, we seem to know precisely what we mean: we think of a political institution such as a ministry or the headquarters of a political party, of companies such as Shell or BMW, or of social profit organizations such as Caritas and Greenpeace. According to the consultancy duo Königswieser and Hillebrand [2], we have understood very little about organizations if we reduce them to buildings or brands. Can an organization be perceived so simply? Are we in an organization when we go through the building's door or when we sign an employment contract? What about when all the employees have gone home in the evening? Have they taken the organization with them in their heads? Or

Kanban Change Leadership: Creating a Culture of Continuous Improvement, First Edition. Klaus Leopold and Siegfried Kaltenecker.

does the organization remain in the rooms, in the furniture, and in the documents, waiting until they are stirred again the next day?

The fact is organizations can be reduced neither to buildings nor to gear-based cog systems, where one simply needs to insert an extra gear to effect an alteration. Organizations are not, to quote Heinz von Foerster [3], "trivial machines" that one can control by pressing a button but rather living social systems with a high degree of complexity. Systems are as dynamic and self-contradictory as the environment within which they define themselves. On the one hand, they are dependent on this environment because it is the source of their supplies (in the form of money, personnel, attention, etc.). On the other hand, they only ever follow their own laws, which, when considered externally, can hardly be identified and therefore influenced only with great difficulty. From the customer's perspective, IBM, Nokia, and Apple operate like classic *black boxes*: no one on the outside can tell how they really work on the inside. However, if these organizations lose their ability to identify what's happening in the market, they run into difficulties.

Organizations always decide how to react to their environment autonomously. Anyone who has ever tried to lodge a complaint against the Austrian Railway or make a personal appointment at a public health office knows a thing or two about this.

However, organizations don't just decide day-to-day issues based on their own laws; they also exclusively build on prior experience when it comes to serious, even life-threatening, change impulses. As we will see in the next example, this can lead to a situation where alarm bells go off across the entire organization and everything is focused on mastering the crisis. "We can't go on like this!" is the exasperated message of the marketing director of a medium-sized organization specializing in software testing. "By now everyone in the marketing team is convinced that our products are no longer cutting edge and that our services are also too expensive." This time, the marketing director's message resonated with the entire executive board. However, instead of falling into hectic activity and kicking off various countermeasures, they took the time for a comprehensive diagnosis: loyal customers with whom the organization had developed a trusting relationship over the years were questioned along with services providers from the organization itself. Firstly, the discussions painted a significantly more critical picture of the organization's operations than they expected: besides the problems mentioned by the marketing director, many customers were dissatisfied with lead times, solution flexibility, and reliability. Secondly, the discussions unearthed very fixed approaches to solutions, for instance, the broadening of the service portfolio through an agile approach; the subsequent development of corresponding know-how; closer collaboration between various specialist departments; and, last but not least, the introduction of Kanban to improve process management in the software development area.

The self-reference of organizations, that is, the systemic stewing in their own juices, can however also lead to resistance against environmental impulses as another case study shows. The business director of a long-established electronics firm seemed to be ignoring the change in the market. Instead of developing online sales, he still relied on personal contacts. In the face of growing competition from large chain stores, he held onto his faith in the regular customers, and in the light of the growing pricing pressure, he relied steadfastly on the "good reputation of the company" and the "special effort that people always make for us." Only once it was clear that the company was up against the wall—despite all faithful customers and special effort—did he give up his resistance to the change. The director resigned and handed the reigns over to his son. As evidence for the worm that was eating away at the organization, the son decided to employ an external crisis manager for the turnaround. And just as symptomatically this manager overestimated the capacity of the organization to change, as he did his own influence. Instead of respecting the individuality of the electronics firm, he worked on the basis of best practices from other areas. Available knowledge wasn't used for the change, neither were employees actively involved. Instead, the son and his crisis manager tried to implement the new company structure top-down. It's no wonder they didn't manage to change course in time. Rather the opposite: some of the most experienced employees were made redundant; others fled by taking sick leave or stuck firmly to their contracts and worked to rule. Like some absurd manifestation of Murphy's law, the most trusted customers started to leave because their most trusted contacts had also left the company. The niche structure planned, which was actually intended to make use of the available specializations, also failed to be established. Finally, after a further 6 months, insolvency was unavoidable.

Would a different concept of change have helped avoid failure? We can't say for sure and we won't speculate. But we can certainly say that the attitude toward change wasn't helpful. Change that hasn't been tried and tested usually has as little impact as pure top-down management: in both cases, there is an absolute lack of respect both for the individual employees and for the organization as a whole. In theoretical terms, there is an absence of awareness of the "precarious nature of the approach to effectively changing an autonomous social system" [4]. This absence necessarily lies at the heart of any professional change initiative. Good intentions are often the exact opposite of good actions.

"Autopoiesis" is the systemic term for organizations that invent, control, and change themselves. It determines how an organization defines its boundaries with respect to its environment, who belongs in it, what products and services are offered, and where it changes, in what form, and when. However, as soon as this change ignores employees and the market, as soon as the corporate value chain is no longer appropriately oriented toward customers' needs, and as soon as competitive needs are ignored or legal parameters disregarded, organizations become acutely aware of this dependence. Typical symptoms here would be drops in turnover figures accompanied by frequent complaints, legislative sanctions, or bad publicity.

SYSTEMS THINKING FOR DUMMIES

"You can't kiss systems," once observed the German organizational theorist Fritz B. Simon [5], pointing out the abstract nature of the phenomenon that is organization. "Systems theory not only doesn't let you kiss her," joked a consultant colleague of ours, "she doesn't even let you read her, let alone understand her!" The polemic cannot easily be dismissed. Systems theory seems unwieldy, academic, and far removed from actual practice, but we nevertheless consider it valuable if we need a suitable understanding of organizations and their possible change. The concepts of complexity, contingency, and conflict are helpful for the right approach. In a nutshell, organizations are thus defined:

- **Complex**, because they consist of many different types of communication networked in various ways and not causally connected with each other. Communication represents the core element of every organization. Consequently, every change management team should expect surprises.
- **Contingent**, because every organization represents a structured combination of such types of communication, although these structures can look very different. This in turn explains why change is at all possible for self-referential systems.
- **Conflictual**, because they're always concerned with pursuing certain possibilities while neglecting others and, in a similar vein, disputing who should decide what is worth keeping or changing and when.

All three concepts are leitmotifs in this book, just like systems thinking in general.

9.1 ORGANIZATIONS CLOSE-UP

But what goes on inside an organization? How does systems theory help us better understand the *black box*? It does this by making provocative claims: the organization is a process focused on communication and decisions rather than people or personalities. Firstly, the organization is not an object, something fixed, or something set in stone. Secondly, the members of an organization are the means and not the center point of the ongoing process of organizing. Because what would happen, asks systems guru Niklas Luhmann, if organizations were to consist of people? If these people were to go to the hairdresser, would a part of the organization be cut off? [6] People don't give themselves over to the organization body and soul but rather based on specific expectations:

- **Expectations regarding content**: "How I want to use my expertise here and contribute to success."
- **Social expectations**: "How I would like to work collaboratively here and receive recognition."
- **Expectations regarding time**: "How I would like to learn and further develop myself."

These expectations give rise to a situation where very specific communication routines and decision patterns are firmly established in every organization. As a marketing director of one media organization stated, "...until the way in which we at this organization speak and act is in our blood." For members, the specific way in which an organization organizes itself becomes something self-evident, indeed even natural. In the words of the marketing director, "This is the way things work around here; forget everything else." Accordingly, these laws of nature are then followed closely, at least as long as survival seems certain. In the light of the average life expectancy of organizations, we should probably add: sometimes even longer than survival is guaranteed.

Must organizations necessarily fail? Is it only a question of time before they collapse through self-reference? Do all experience the same destiny? Not at all, as shown by many organizations, some are more than 100 years old and still agile. The same can easily be argued from a systems theory point of view. Ultimately, for all their self-made laws, organizations are not doomed to a total lack of communication with the outside world:

- Just like any living system, organizations act according to their **learned patterns**, but they are also **open to the environment around them**.
- They follow their own policies, but they must react to **new policies in the market** with timely **internal policy changes** in order not to find themselves left out in the cold.
- They **shut themselves off from most external impulses** but then react to particular things very sensitively.

As illustrated in Figure 9.1, change management should take the strategic openness of organizations into account. However, the right conditions should be met in order to mobilize sufficient change energy. A sense of urgency must be advocated; otherwise, there is the risk that the strategic window of opportunity will simply be shut again.

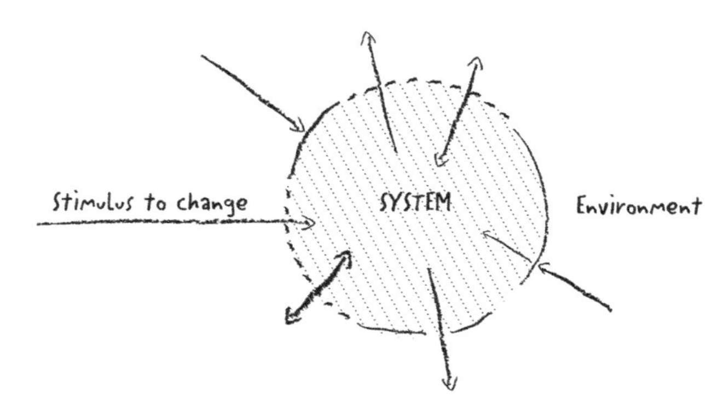

FIGURE 9.1 A system and its environment.

Certain indicators for the management of change can be extracted if we follow this systemic view of organizations. The systems perspective immediately makes it clear that organizations don't function like machines. Consequently, changing them takes more than the usual toolbox.

The fact is that the hierarchical bureaucratic super companies of the twentieth century are increasingly losing ground to a multitude of heterarchical, network-like organizational forms.

If you are prepared to take the risk of exaggerated polarization, two completely different organizational models can be outlined:

1. A **mechanistic** model based on linear cause-and-effect relationships, rational plans, and central forms of control.
2. A **systemic** model based on various feedback loops, contradictions that are not totally transparent, and latent uncontrollability.

While the established processes in the first model are primarily based on purpose-driven approaches, hard facts, integral instructions, and controls as described in Table 9.1, in the second model, it is a matter of making sense, integrating hard and soft factors, and faith in self-control. In mechanistic thinking, change is a matter of instruction and direction; in systemic thinking, it is one of dialog and conviction.

TABLE 9.1 Organizational images

Mechanistic organizational image	Systemic organizational image
Manageable and clearly structured	Complex and self-contradictory
Linear cause-and-effect relationships	Various knock-on effects
Centrally controlled, leading according to a rational plan	Self-controlled, following its own laws
Formal logic	Integration of contradictions
Primarily purpose driven	Primarily meaning driven
Hard facts and rational relationships	Hard and soft factors, emotions
Structure and process oriented	Oriented around patterns of behavior and routines
High significance of individual parameters and controls	High significance of shared reflection, collaboration, and systemic results
Change through instruction and command	Change through dialog and conviction
Central leadership	Distributed leadership

Furthermore, the systemic perspective makes it clear that environmental changes always constitute an unreasonable demand for existing organizations. Despite all official innovation rhetorics, social systems are conservative per se—they are ultimately geared for continued existence. Of course, from a systemic point of view, a high degree of ignorance is at the heart of any organization. Things are perceived if they want to be perceived. If something doesn't appear on the radar of perception, then it will seldom become an issue for the organization's internal communication. Considered in this light, the disinterest of many organizations in environmental problems—including, but not limited to, ecological problems—isn't hugely surprising.

The systemic model makes it clear that organizational changes must always be organized from the outside in.

Every market situation must be perceived correctly and approached with compelling products and services. Tailor-made offers must be brought to the market without unnecessary delays; they must be tested and, when necessary, modified quickly. Change is thus to be driven both by need and value added, which, as we will see, can lead to strikingly different change options.

If development impulses are perceived from the outside, people in the organization will initially ask about the need for change. Do we really need to? Is it so important? Is it really important and urgent? If the answer is yes, then we need clarification as to how comprehensive the necessary change is. Where is the emergency situation we need to avert? What risk is there for the organization if it doesn't change? And what type of change does the organization need in order to successfully manage this risk? "Why the great hoo-ha?" the senior director of a family-owned enterprise in the pharmaceutical industry wanted to know at the beginning of the initial discussion to which his two sons, junior directors, had invited us. The senior director could just as little understand the change in their customer base as perceived by the junior directors as he could their fear that one of the organization's sectors had been going down a dead end for a long time. And of course, it irritated the senior director that two external consultants had been invited, who themselves felt rather a lot like uninvited guests at a Sunday family lunch.

Questions about the capacity for change followed the questions about the need for change. "What do we need in order to create a new strategic alignment?" asked one of the junior directors, concerned. "Which of our core competencies can we use and what do we need to develop as quickly as possible?" The ensuing discussion can be considered typical of the beginning of a change initiative. While the one son immediately wants to go *in medias res* and define concrete measures and the other son yearns for a precise diagnosis, the senior director remains doubtful as to the need for this whole change hoo-ha. A thrill-packed discussion revolves around the "for" and "against" points of the various options. What are the arguments for retaining the status quo? What are the arguments for change? What change is necessary? And how can this be executed with the available resources?

9.2 A ROADMAP FOR CHANGE

By combining environmental demands and system competencies, organizational consultants Barbara Heitger and Alexander Doujak have developed a roadmap for change management [7]. Something that's called change is not yet anywhere close to *being* change, at least not the same kind. The strategic realignment in the example

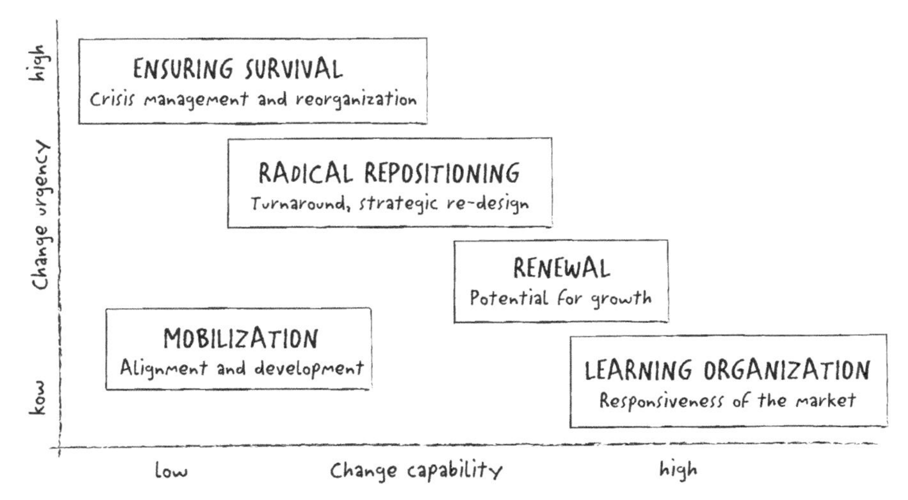

FIGURE 9.2 Roadmap of change management.

earlier requires a different change process to that of the implementation of an organization-wide leadership program or the conversion to agile software development. Furthermore, structural factors such as the organization's history, size, branches, structure of ownership, and last but not least past experience of change projects play an important role.

As Figure 9.2 shows, through the combination of the varying degrees of urgency for change (vertical axis) and the capacity to change (horizontal axis), five types of change can be identified:

1. **Ensuring survival**: If the urgency for change is so great that we are first and foremost concerned with ensuring the survival of the company and correspondingly tough steps are recommended.
2. **Mobilization**: If the pressure for change is relatively low and one is currently sailing on calm seas. Well-timed learning and the highest possible reaction time are recommended to avoid being surprised by sudden gusts. The perceived need for change often comes into play in this situation.
3. **Radical repositioning** is used, given the perceived need for change, to prevent a possible crisis by getting the organization to realign itself strategically and develop its change competence.
4. **Renewal** is an approach whereby one attempts to test out new maneuvers in the case of moderate pressure for change. Investment in innovation is intentionally made, and the change capability is augmented.
5. **A learning organization** perceives change as a part of the daily business life in which mental models are shared and strategic meaning is applied to learning.

In a nutshell, the change roadmap defines the area of conflict between more radical, rather revolutionary change options (upper left) and emerging, evolutionary

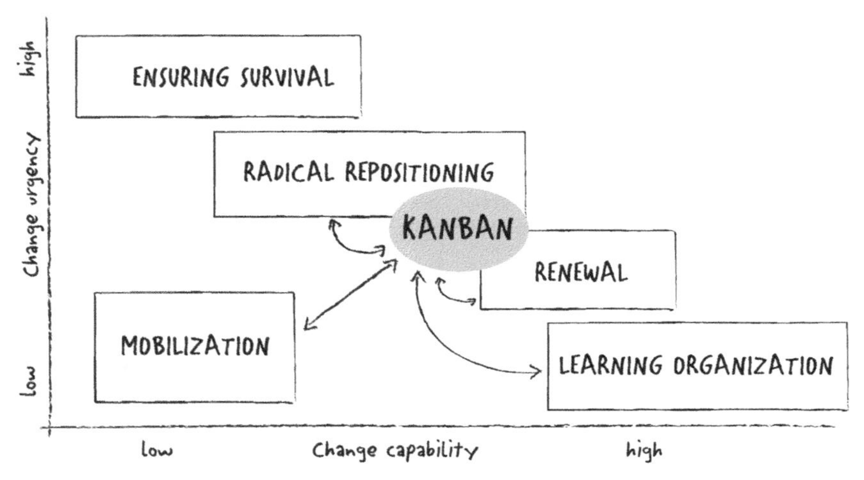

FIGURE 9.3 Kanban on the change roadmap.

initiatives (lower right). While the former approach includes deep incisions, the latter one tries to drive change by continuous improvement.

Figure 9.3 illustrates the position, in our opinion, of Kanban on the roadmap. Kanban normally advocates a low to medium change pressure and doesn't demand a particularly high change capability of an organization. The arrows underline the strategic agility of Kanban, which can contribute to mobilization as well as repositioning. Continuous improvement encourages successive adaption and growth at the same time. Furthermore, Kaizen thinking pulls Kanban in the direction of a learning organization that is above all concerned with better customer and service orientation.

Regardless of where you place Kanban on the map, a purely analytic application of the change roadmap is somewhat evocative of a bill that is made up behind the landlord's back. In other words, you mustn't forget the people who achieve the changes—or don't. We're concerned here with how individuals set themselves in motion, what motivates them, and how these motivations are processed within the system. The number-driven logic of change management is inseparable from the logic of feelings—and this is what we'll look at in the next couple of chapters.

WHAT YOU CAN TAKE AWAY FROM THIS CHAPTER

In order to be able to shape change processes professionally, it is not only necessary to have an appropriate understanding of the external drivers of change, but it is far more important to clarify the internal processes of the organization.

Systems theory helps us to rethink organizations in terms other than the mechanistic mold. They can rather be understood as living social systems that are highly complex and self-contradictory. On the one hand, they are dependent on the environmental systems that support them (with money, personnel, recognition, attention, etc.), and on the other hand, they ultimately follow their own laws, are autonomous, and are self-referential. In a nutshell, organizations don't let anybody tell them who's boss.

Their reactions to change impulses from their environment are accordingly stubborn. Most impulses are ignored, a further portion falls victim to the established routines of perception and communication, while only a tiny percentage is actually taken seriously.

The combination of questions about the need and the capability for change allows us to distinguish between different types of change management: crisis management to ensure survival, radical transformation, strategic innovation, mobilization initiatives, or learning organizations.

Kanban change management can be positioned approximately at the center of this roadmap, but it also boasts a high degree of strategic flexibility.

10

ORGANIZATIONAL AND PERSONAL CHANGE

The German organizational consultants Klaus Doppler and Christoph Lauterburg have convincingly shown how personally people take systemic demands. "Do I need to do this? Can I do this? Do I want to do this?" These are according to their landmark book the questions every person affected by an organizational process of change ask themselves [1].

Do I Need to Do This?

The general question concerning the urgency or at times even the indispensability of the change is motivated by the following questions:

- Why change?
- What are the causes for it?
- Is management telling us the whole story or are they keeping some of it to themselves?
- Is it really so important—or might there be more urgent problems that we should be taking care of?

It is hardly surprising that trust is a critical element in processing these questions.

As we will discuss more closely later on, these seemingly neutral questions immediately indicate the quality of the various relationships: in the organization as a whole,

Kanban Change Leadership: Creating a Culture of Continuous Improvement,
First Edition. Klaus Leopold and Siegfried Kaltenecker.

in individual teams, between the management and the staff, and also between the employees themselves. "How open is discussion in our organization?" we were once asked right at the beginning of an interview. What the senior developer—described by the head of development (my actual client) of an international insurance group as one of my "absolute star performers"—told me might well have displeased the head of development. "I don't have any faith in him (the head of development) anymore," said the developer, giving vent to his dissatisfaction while proceeding to reply to our further questions. "He just tells stories. Critical feedback is not welcome. Officially of course we are all integrated and to this end we have nice little information events. But our suggestions aren't listened to. Apparently, our own experience has no role in the change process. You have the feeling it's all already decided and we're just there to nod."

Can I Do This?

Regardless of how the need for change is communicated, on the individual level, the question "Do I need to do this?" is intimately related to the question "Can I even do it?"

- Am I able to cope with what is heading my way?
- Do I have all the skills I need to master the change?
- What are my chances for good results?
- What counts as success under the new conditions?

Trust also plays a critical role in personal "changeability" [2]: on the one hand, the trust of others in one's own strengths and, on the other hand, trust in oneself. After all, what we're concerned with here is an honest appraisal of what one is prepared to do in the course of change. In this sense, it's not so much to do with private feelings as in "Am I about to break up with my partner?" or "Have I just fallen in love?" or personality structures such as "Is my nature that of someone who welcomes change or am I conservative; am I open-minded or fearful when it comes to change?" Beyond life situations and psychographics, corporate culture plays an overwhelming role in one's own perception of competence.

The following was said in a group discussion with the management of an IT service provider: "To be honest, I don't really feel like we enjoy a huge amount of respect in our company," to general nodding. "Do you have the impression you're a valuable part of this organization?" This resulted in questioning looks and a collective shaking of heads. "Are your contributions really valued or is it perhaps taken for granted that everyone will bend over backward for the company?" A second manager spontaneously said, "Above all, when it's really critical, we're left in the lurch! Do you even feel like you get any support?" A third manager, laughing bitterly, said, "Support? They don't even talk to us; they just hand out orders!"

Do I Want It?

Anyone who has ever been in a similar situation will easily understand how quickly, indeed instantaneously, the previously described organizational experiences are projected. It is this emotional calculation more than anything else that dominates the willingness to change:

- What's in it for me?
- Are the new activities interesting?
- Who will I be involved with?
- Is there a risk of losing something: money, relationships, interesting career prospects?
- Can I expect to gain something from the change?

The answers and their associated emotional qualities determine our fundamental attitude to change. The amount of energy that can be mobilized depends on this attitude. Do I have a positive attitude concerning this change or is it actually negative? What is my individual stand with regard to the information I have received, the discussions I've had, and the estimations I've made? What do I think? What's my gut feeling? And how will I act accordingly?

10.1 THE ICEBERG OF CHANGE

The sum of my answers to "must," "can," and "want" determines my attitude toward change. The "iceberg of change" illustrated in Figure 10.1 should remind you that it's only partly about matter-of-fact reasons.

> *Regardless of whether these reasons make up exactly the much-discussed one-seventh of the entire change event, one thing is certain: under the rational tip lurks the emotional volume of the iceberg, and this large part lies beneath the waterline of pure reason.*

The image makes explicit the fact that in terms of the change itself, the justifications for the change, the communication of the strategies and goals, and the representation of a project plan normally constitute but a fraction of what actually takes place in change phases. That many change initiatives end up wrecked by the iceberg is related to the fact that these initiatives are often only concerned with what lies at the surface of the water. Additionally, the orientation is often further limited due to bad communication between the captain, officers, and shipmates. However, if an initiative only takes the notorious tip of the iceberg into account, then it should come as no surprise if the initiative does a Titanic.

Nevertheless, one would hope that the emotion-oriented portion of the iceberg doesn't remain hidden forever and that it's not only criticism and complaints that

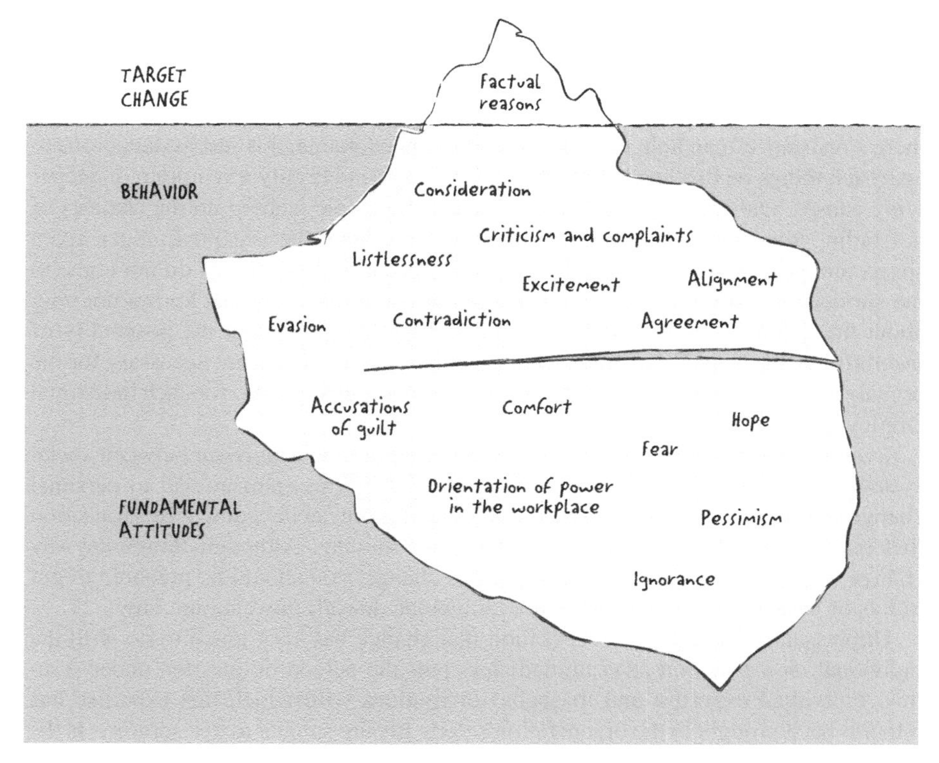

FIGURE 10.1 The iceberg of change.

appear but also curiosity, interest, and engagement. Everyone's aboard, the course has been set, and each gets to work so that it's full steam ahead with the changes. A good sense of humor is useful to banish the change monster and free up new energies.

THE CHANGE MONSTER

"You have to expect the unexpected. Unforeseen trials and tribulations will occur…," prophesies Jeanie L. Duck in her highly recommended book *The Change Monster* [3]. Monsters are also the order of the day in Disney's animation film *Monsters, Inc.* (2001). In this film, monsters must systematically frighten children because the company they work for uses the children's screams to generate energy. The monsters go about this gloomy business until 1 day they happen to discover that laughter produces significantly more energy than screams. This pleasure independently creates completely new relationships. From this point on, the monsters are glad to be this close to the children. Consequently, Monstropolis changes its strategy from that of making children scream to making them laugh.

Unfortunately, a positive flow of energy at the beginning of a change initiative is the exception rather than the rule. Normally, people start to have concerns early on but skepticism is not usually voiced. Instead, people make themselves scarce, avoid contact, or display indifferent. However much the formula "wherever there's change, there's resistance" can help us understand these phenomena, it is just as dangerous to approach things on this level. To fight, flee, or play dead is only a symptom of deeper lying causes. Only when we begin to measure the whole iceberg do the chances of not failing sooner or later begin to increase. Fear must be accepted, anger given space, and pessimism respected. It is precisely because these causes do not come to the surface so easily that we need professional communication that knows not only about the depth of the sea containing the iceberg but also about the possibility of insightful dives. Depending on the change situation, these dives are necessary for the organization as a whole, for every team involved in the change, and for each individual employee.

Even if systems gurus don't get bored pointing out the difference between social systems and psychological systems, the transition from organizational to personal change challenge may be seen as a central issue. It is only at the point of this transition that real commitment evolves and things start moving. Although things anyway change with time, if you want to control this change, you should be prepared to put in lots of hard work yourself. After all, faith alone doesn't move mountains.

Unpleasantly enough, the recognition that change has very much to do with the individual *on a personal level* immediately puts the person in question under scrutiny. Individual expertise and the behavior routines with which this expertise has hitherto been brought to the organization's daily life are subject to this scrutiny. Is the person's knowledge still relevant? Does their competence contribute to the success of the organization as a whole? Is the form in which this contribution is made still adequate? A great deal of uncertainty accompanies these sorts of questions. In the context of the change process, you must necessarily ask yourself what this process will mean for your own status and for your network of relationships within the organization.

In the style of Henry Ford's famous complaint that every time he asks for a pair of hands they come with a brain attached, a huge amount of emotional energy is pumped into the supposedly rational system of the organization. Things sometimes spill over unexpectedly. A department manager of a bank recently asked us for support. The bank was obviously struggling to control its emotional pump. The fact that the scales were tipped particularly high in the department that handled international business was hardly surprising given the daily business news. However, the fact that in one of the hitherto most successful teams the nerves were completely shot was a sudden surprising escalation. "Catfighting," we were succinctly told by the leader of the investment department that had been very successful over the years after we asked him for his thoughts. On the one hand, the department manager's explanation was interesting because the team affected—with which we were due to carry out a strategy workshop—consisted only of men. On the other hand, the manager's visibly negative evaluation led us to suspect a deeper lying problem. You don't have to be a riddle master to make the link between this problem and the fate of those who had suddenly gone from hero to zero. "Windbag," "figurehead," and "chair warmer" were but some of the names the team members called their departmental manager during our initial meetings. The aggression with which the team members did down their boss and other colleagues took our breath away. With each new conversation, we became more convinced that the planned 1-day workshop would only scratch the surface of the problem. As we then reported back our most significant impressions to the departmental manager, the emotional pump provided the next spillover. With a bright red face, the departmental manager responded that we "obviously weren't willing to solve the problem in the allotted time," and he therefore felt obliged to release us from the contracted work.

What have we gained from this story?

- The **confirmation** that the iceberg isn't just an academic model.
- The **insight** that even as a consultant you can easily sink in the sea of feelings.
- The **lesson** that with change, time is always a factor.
- The **proof** of a well-known dilemma associated with the factor of time: on the one hand, nobody will dispute that change needs time—in our experience, the more emotion involved, the longer the required time—and on the other hand, nobody is overly happy that the time frame for the successful implementation of change is often hard to estimate.

10.2 THE CHANGE CURVE

While the iceberg illustrates the interplay between content-rich *hard facts* and social *soft facts*, the so-called change curve is often used to better estimate the time factor. The curve in Figure 10.2 attempts to make the normal progression of personal change management comprehensible.

The change curve was developed by the American psychiatrist Elisabeth Kübler-Ross in the 1960s and has since been used in a variety of ways [4]. It is one of the classical attempts to inject a certain amount of structure into the systemic incalculability of change. The dynamic of personal change processes is established with the help of two axes: the time axis and the axis of one's own feeling of competence. The graph relies on the tendency of many people to react to change with initial shock—at least that change coming from external sources as in the case of unexpected organizational change, acute crises, or an intensive need for reorganization. We are shocked, we are surprised, and we feel ourselves caught off guard. These experiences are accompanied by a loss of one's own feeling of competency. This is a key point of the curve model. It instantly becomes clear that:

- **Something must have happened** that has perhaps been overlooked for a long time.
- **It clearly can't go on** as it has been.
- A **multitude** of questions are waiting to be asked.

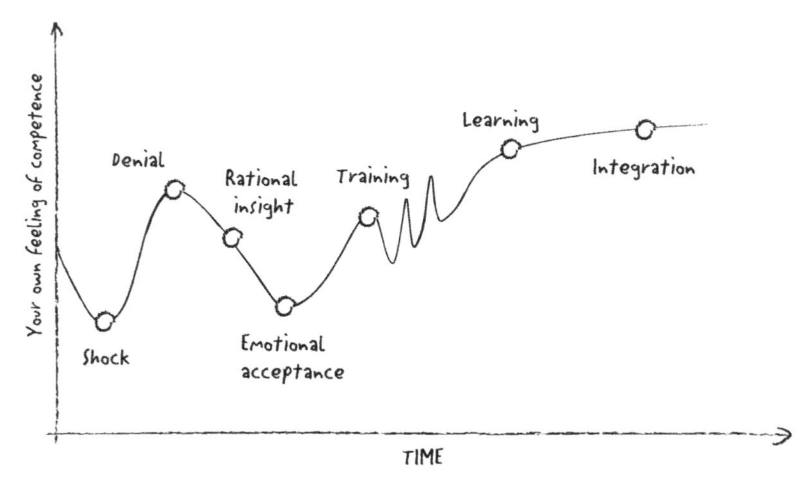

FIGURE 10.2 The change curve.

- **Feelings of uncertainty, threat, and irritation** are getting the upper hand.
- There is **no quick fix** that will banish these unpleasant feelings in the case of high pressure for change.
- The **change challenge cannot be ignored** while you get on with the daily business.
- You yourself **are not able** to drive the change.

Naturally, change is experienced differently by everyone. What is essentially a new opportunity for the first person goes completely against the grain for the second, makes the third nervous, and satisfies the fourth. And of course, there are organizational cultures constantly seeking new challenges. Nevertheless, the shock triggered by change procedures, be it the depth of the shock, the vehemence of the rejection, or the duration of the repression, never ceases to amaze us.

Viewed from a psychological perspective, systemic shock is accompanied by a personal hurt. What has hitherto been achieved is no longer enough; one's own technical expertise is perhaps no longer in demand; personal commitment is seemingly worthless.

Even relatively mild "mobilization" or "renewal" change can trigger a strong headwind: management doesn't want to participate in the training; teams don't bother with retrospectives; senior management cancels participation in the steering committee for the *n*th time; and so on and so forth.

10.2.1 Fear and Resistance

How can we explain such symptoms? Why do so many people find it difficult to see change as a chance? What causes resistance? As our experience shows, changes trigger a multitude of fears:

- That **one's own uncertainty becomes visible**
- That **skeletons come out of the closet** and criticism, shortcomings, and deficits move to the forefront
- Of being insufficient in the face of **the complexity of change management**
- That the **change won't bring improvements** but rather have negative consequences

Overwhelmed by fears of this sort, self-esteem is susceptible to being washed away along with the individual sense of competence. It is exactly this dynamic that the change curve tries to explain. The following rule of thumb applies: the more comprehensive the change challenge (up to possible firing), the larger the shock (i.e., threat to competence) and the deeper the hurt (i.e., the perceived loss of self-esteem). The probable consequences of such a dynamic are easy to see: exchanging accusations, denial, trivialization, humiliation—in brief, resistance to change.

This resistance can take on many forms: radical or subtle, short-term choleric or long-term unmotivated, outspoken, or silent. Whatever the type, the resistance mostly serves at an emotional level to revamp our injured self-esteem. To put it simply, if we deflect the change challenge from ourselves, then we're back on top!

We'll never forget how we were confronted with this particular form of self-empowerment in a factory where we were to facilitate a retreat on "new processes." As we started our planned on-site inspection to get to know the work relationships and chat with the participating shift supervisors on the premises, we immediately felt as though we'd walked into a picture book on the subject of resistance: folded arms, frowns, heads shaking, rolling eyes, and whispered exchanges. "Looks like their lordships have had some new ideas" was the soundtrack to these gestures. "Maybe they should do their own homework for once." "They know where they can stick their new processes." Or "Retreat? That's the biggest pile of rubbish I've ever heard!" A modicum of politeness stopped them throwing us out of the building headfirst. What we went on to experience while wading through resistance seemed to be perfect proof for the classic logic of failure: the change was experienced as coming from the top down; it was enforced without any consultation and was at times even perceived as intentional chicanery.

10.2.2 Rational Insight and Emotional Acceptance

In the course of the retreat that had finally been accepted by all, the major topic of conversation was that there were practical reasons for the organizational change: from production bottlenecks to chronic machine breakdown and bad shift handover. However, the initial emotional resistance in the open discussion made it possible to talk about the required changes in a rational way. In the course of the retreat, we were repeatedly reminded of the typical behavior of the change curve. Regardless of how resistance is justified, we only begin to accept the necessity of change emotionally—as the third and fourth stages of the typical roller-coaster ride of change clearly show—if we engage with the situation in a more sophisticated way.

In terms of change management, we would like to establish in advance that suitable communication opportunities and their professional facilitation are critical to success in this phase. As the retreat just described shows, a careful shaping of such communication opportunities can drastically reduce the shock phase. Examinations, individual coaching sessions, team workshops, and large group events cannot just magic the change challenge away. They help process this challenge better.

Helping people help themselves is even more important due to the fact that an individual's perceptions of their own competence have reached a low point in this section of the change curve.

At the same time, this phase reinforces the systemic Münchhausen principle very well. We drag ourselves out of the swamp because—a propos the principle of self-control—nobody else can do it for us. The retreat clearly shows how this can be reinforced through shared conflict management forums. Small groups made it possible to bring the extant displeasure out in the open. The most important points of criticism were written on cards and presented in the plenum. After a consolidation of the most important themes and their shared prioritizations, new groups had to come up with possible improvement measures. These measures were presented in the form of a gallery and again discussed openly. The entire group then created a fixed implementation plan that ultimately won the approval of the Head of Production as well as that of each shift supervisor.

10.2.3 Applying the New Behaviors

Of course, such events must not be overestimated. Even when the right people are brought together and provided with good tools, these events are no miracle change management cure. Measured out in homeopathic doses however, they can be very healing indeed. In addition to this, retreats, workshops, and training programs represent good opportunities for practice. As the fifth phase of the change curve demonstrates, these opportunities are essential for processing the change productively. Ultimately, during the process of change, it is almost always the case that something new must be learned and, as is all too frequently forgotten, something old must be unlearned. Structures must be brought to life; processes practiced; competences acquired; patterns of behavior cultivated. A culture of open conversations and clear decision-making is therefore even more important during this phase since social systems make sense primarily via their mode of communication. Faithful supervision helps to make the unavoidable to-ing and fro-ing during the learning process as constructive as possible.

Such guidance is even more important since a sense of competence also plays a role during this phase. High in the sky 1 day, the pits the next—that's what the emotional aspect of the learning process for this phase can be like. It is a phase in which doubt and strength, confidence and skepticism, and joy and anger are never far apart. The merit of reaching milestones, the positive effect of quick wins, and particularly appreciating successes are all tried and tested ways of supporting sustainable change.

10.2.4 Learning and Integration

As the curve further proceeds, the practice phase is followed by the phase of multifaceted learning. Here is where we realize that:

- The **change is really doable**.
- Things can really **be different** to how they were before the change.
- Many **questions are answered satisfactorily**.
- People are dominated by **positive emotions**.
- People **know once more** how things work.
- The change makes sense, above all with regard to **current and future certainty**.
- Together with all the other learning steps, we are increasingly.

These results lead us to the final section of the change curve: the so-called integration phase. Here, we are above all concerned with stabilization, that is, establishing new routines and patterns, which in today's world is even harder to achieve since the next change wave is most likely just around the corner—unless, due to the simultaneity of different change initiatives, it has already arrived.

In actual fact—and this is the downside of an ideal model—change in organizations doesn't only follow one single curve:

1. The graph takes on a specific shape for each person. As already observed, many factors contribute to the experience of shock, vehemence of the rejection, difficulty of acceptance and practice, and, last but not least, duration of the integration.
2. Experience shows that the change curve is also influenced to an extent by hierarchical positions. As Figure 10.3 demonstrates, managers often enter the various phases earlier than employees, which can be attributed partly to earlier information and partly to more senior know-how.
3. The curves differ because certain organizational sectors are affected by change to different extents. Depending on organizational complexity, the introduction of agile processes in software development, for example, will affect the sales and marketing department very differently to the project management or HR department. The merger of two cash management departments in the process of a bank fusion will put much more pressure on the employees in this sector than the still independent department of international finance. And the market-driven installation of new machinery will be more demanding for the shift supervisor in production than for the marketing management.
4. It's pure madness to think that personal change processes evolve in a linear, progressive manner. In reality, such processes are almost always determined by incalculable disruptions, to-and-fro swinging, or so-called laps of honor: management holds onto the rhetoric of change; teams evolve in the stipulated direction only to get stuck halfway, as it were; and employees develop new forms of behavior, capitulate to pressure, and revert to old patterns.

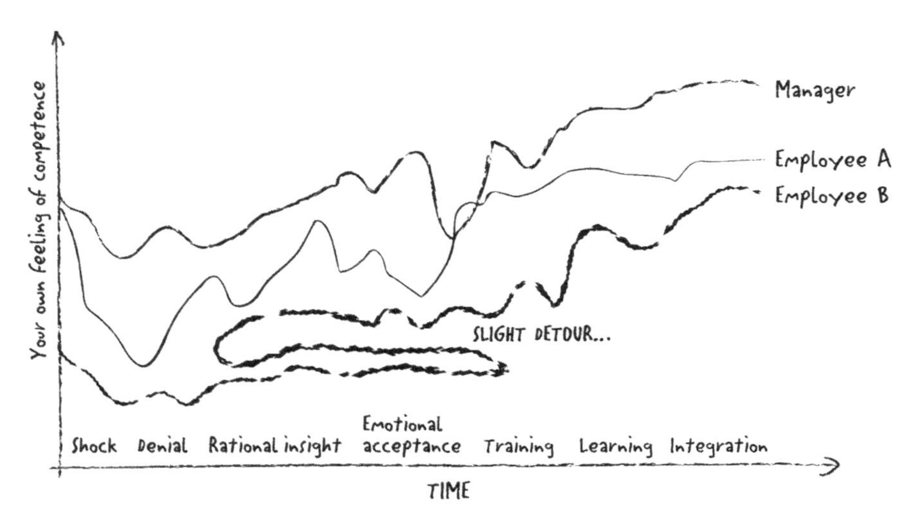

FIGURE 10.3 Various change curves in an organization.

So why do we nevertheless value the change curve? The model helps, we believe, to perceive change not as a static block but rather as a dynamic combination of various phases. It underlines the intimate relationship between the factual justification and the emotional processing of change and shows that successful change management is dependent upon a professional approach to the resistance, uncertainties, doubts, and fears that materialize.

"I know this all sounds very complicated," said the Austrian ex-Chancellor Fred Sinowatz in 1983 when he addressed the situation in his country, which involved countless scandals. In his governmental address, Sinowatz highlighted the various problematic areas of the political change that were, in his view, completely necessary. In so doing, he emphasized the complexity of the change processes before falling victim to several of them himself. However, for both political and organizational change, it is mandatory that the actual circumstances be addressed:

- The **actual challenge**
- The **organizational situation** in which this challenge is addressed
- The **management** that has strong influence on this work process
- Each **individual** that shapes this process according to personal policies

The shape of the change curve as a result confirms two truisms of change management: firstly, that all roadmaps and models are fundamentally false, although some, as the "expedition into uncertainty" shows, can be useful nevertheless; and secondly, that models are helpful as long as you don't believe in them. Change managers and team members should therefore be wary of treating the change curve as valid for all situations and under all circumstances. They should also be wary of underestimating different phases of change and their emotional intensity. Armed with this wisdom, then, we will now embark upon one further expedition into the "logic of feelings."

AN EXPEDITION INTO UNCERTAINTY

In his book *Sensemaking in Organizations* [5], Karl E. Weick tells the story of a Hungarian military unit carrying out a maneuver in the Swiss Alps. Surprised by a storm, this unit went missing for 2 days in the snow and ice and all hope seemed to be lost. However, on the third day, they returned, uninjured, to their base camp.

How did the unit manage to extricate itself from this hopeless situation? The unit commander explained that after they had given up all hope, one of the soldiers incidentally found a map in his rucksack. Everyone instantly relaxed, struck camp, and survived the snowstorm. The next day, the commander led his troops back to base camp with the help of the map.

It was quite a surprise when they later discovered that the map didn't show the Swiss Alps but instead the Pyrenees.

WHAT YOU CAN TAKE AWAY FROM THIS CHAPTER

"Do I need to do this? Can I do it? Do I want to do it?" These are the questions that every person affected by the change process asks.

The transfer of the general, organizational change to an individual, personal challenge is central to professional change management.

The iceberg model demonstrates that the rational justification of the need for change is often only the tip of a change event that emphasizes the emotional volume. We are less interested in the symptoms of this event than their more deep-lying causes.

As opposed to this, the model of the change curve shows that a change process consists of various phases that inseparably connect logic and emotions with each other.

As long as you do not confuse these simplified roadmaps with the complex territory of a change process, they can prove of great use in modeling change management.

11

EMOTIONS IN CHANGE PROCESSES

Regardless of the model for change management, change is inseparably linked with intense feelings. Even the etymological root of the word "emotion" supports this link: "emotion" derives from the Latin word *movere*, "to move." But what is it that *moves* in change processes? What feelings are stirred? And how can they be used to achieve the desired change?

> *Emotions are often associated with images of chaos, raw energy, and uncontrolled dynamics. Consequently, they appear to be diametrically opposed to the rational concept of organization.*

However, in his theory of the affect–logic, Swiss psychoanalyst Luc Ciompi convincingly established that feelings and thinking are by no means separate worlds [1]; they are rather in constant contact with each other as also demonstrated in contemporary neuroscience. According to Ciompi, thinking that is devoid of feelings doesn't even exist. Affect–logic is defined as a comprehensive body–soul condition that can have different levels of intensity. Such emotional–cognitive states can appear consciously or unconsciously and can last anything from a few seconds to many hours or even days, for example, in manic or depressive moods. Feelings are also psychosomatic phenomena as in the following sayings: "my heart skipped a beat," "a chill ran down my spine," and "venting spleen." Affect–logic drives "energetic conditions, or, more precisely: patterns of energy distribution" [1, p. 23] that constantly influence our focus of attention.

Kanban Change Leadership: Creating a Culture of Continuous Improvement,
First Edition. Klaus Leopold and Siegfried Kaltenecker.

It's clear as day that these energetic focuses have a central role in the phases of change. In their book *Managing Cuts and New Growth: An Innovative Approach to Change Management* [2], the consultancy duo Barbara Heitger and Alexander Doujak pick Ciompi's theses up and apply them to the area of organizational change management. What are the characteristic feelings that appear during change? What is their specific function? What do they bring about? And how should we deal with them?

Heitger and Doujak stipulate four categories of feelings that typically appear in change processes:

1. **Uncertainty, worry, and anxiety** appear first and foremost in the first phase of the change curve.
2. **Anger and aggression** determine the second phase of a typical change flow.
3. **Sadness and disappointment** are distinguishing features of the "emotional acceptance" phase.
4. **Enthusiasm, joy, and courage** accompany recognition, intensive practice, and integration in change processes.

We will next describe these categories in further detail.

11.1 UNCERTAINTY, WORRY, AND ANXIETY

The main function of the emotions in this category is to accumulate energy and concentrate on possible danger zones. While uncertainty does this in a diffuse manner—along the lines of "I see that I'm drifting but *where* am I drifting to?"—worry is more focused. We worry about someone or something in particular: "I am drifting because of where I was but would like to know where it's taking me."

Fear fuels the feelings of worry and uncertainty. "Help, I'm drifting and that means I could drown!" Even when fear constrains us internally, it also has a positive side, which is that it can help identify present threats more precisely and assist us in readying our defenses. "Which wave is breaking on me right now? And where is the nearest life vest?" One classic reaction here is flight: "Just get me out of here!" However, that's easier said than done if you just happen to be in the middle of the ocean. The search for possible life vests simply becomes more urgent.

In practice, such feelings do not appear sequentially. It's not possible in day-to-day life to clarify whether it is uncertainty, worry, or fear that we are feeling. To top it all, these feelings are often concealed. Since uncertainty and fear make people vulnerable, these feelings are often presented in a raw shell of aggression, particularly in men. Accusations and degrading comments are symptoms, just as in the case of the shift supervisor during the retreat described in the previous chapter. These feelings are also concealed by being repressed, as evidenced by many of the participating shift supervisors who intentionally waited around in the background to see what direction the event would take. "Will everything calm down again in a bit? Will the thing I was worried about sort itself out on its own? Are my biggest concerns going to take care of themselves?"

Alongside struggle and aggression, fleeing from and ignoring them are common ways of dealing with unpleasant feelings. In daily business, "fleeing" doesn't just refer to radical measures such as resignation or illness; "work-to-rule" policies and making scapegoats of others are also typical responses. Ignoring, willfully not listening, eavesdropping, trivialization, and blocking the change challenge are further prevalent mechanisms.

The American organizational scientist Edgar H. Schein has shown that in change initiatives two types of anxiety dominate: survival anxiety, as we have already seen, and also learning anxiety [3].

Both anxieties can originate from very different sources. In many situations, survival anxiety is not necessarily concerned with economic survival in the sense of "tomorrow, I'll be sleeping under a bridge"; rather, it could be fueled by:

- The threat of **loss of status**: "Tomorrow, I won't be a manager anymore!"
- The **devaluing** of your own expertise: "Tomorrow, all my experience as project manager will be worthless!"
- The **threat of dissolution** of a well-known environment: "Tomorrow, I'll be working with a completely new team!"

In the context of learning, anxieties are instilled through both the necessary acquisition of new skills and knowledge and the equally necessary unlearning of old patterns. For example, anxiety about:

- Temporary or permanent **incompetence**: "I simply can't do that!"
- Having to expect **punishments or at least sanctions** due to the incompetence: "If I don't manage it, I'll lose my position!"
- Suffering a personal **loss of identity**: "I've been a software developer all my life. Why do I suddenly need to analyze and test as well?"
- **No longer being a member** of a particular group or community: "What if I suddenly lose the connection to my colleagues?"

DEAD HORSES

Apropos necessary unlearning, American management expert Gary Hamel very succinctly says:

> Dakota tribal wisdom says that when on a dead horse, the best strategy is to dismount. Of course, there are other strategies. You can change riders. You can get a committee to study the dead horse. You can benchmark how other companies ride dead horses. You can declare that it's cheaper to feed a dead horse. You can harness several dead horses together. But after you have tried all these things, you're still going to have to dismount [4].

As has already been mentioned, we're not only concerned with learning new things during change processes. We also need to unlearn all that might stand in the way of the change. Let's refer to Gary Hamel's piece of Dakota tribal wisdom that concludes that you always need to dismount from a dead horse if you want to go further. In organizational terms, this means that outdated routines must be unlearned, concentration points that are no longer important must be removed, and finally, old forms of behavior must be overcome in order for the change to have a chance. This is a process that requires time, patience, logic, and above all taking a deep breath. Ultimately, as Mark Twain obviously knew, you cannot simply throw all old habits out of the window. They must be carried down the stairs, step-by-step.

This is an energy-consuming business. It's no wonder that the mucking out required for any radical change provokes significant resistance. Here, we are reminded of a workshop with IT project managers about the introduction of agile software development using Scrum. The message from the external Scrum expert raised a few eyebrows: project management using the waterfall model was apparently dysfunctional; only the agile approach generated *business value* according to him. Everyone could forget their plans and certainly the idea of a traditional project manager.

It's hardly surprising that we needed to rescue the expert from crucifixion. Equally unsurprising was that his fate was sealed because he questioned everything that the project managers had taken as gospel for so many years. Or worse, the devaluing of the previous experience was accompanied by an emotional suffering that generated resistance to the plan. "Just what is he thinking?" one of the project managers said in the break, summing things up. "That we've produced only crap up until now?"

As in the case study described, the psychodynamic of unlearning is ignored in many change projects. Instead, people assume that change in and of itself is enough to motivate people to learn.

> *The anxiety of feeling at least temporarily incompetent in the processes of learning and unlearning is not given any importance, while people concurrently struggle with supposed technical problems.*

One of the greatest challenges in dealing with such existential feelings is accepting them. This acceptance must begin with the individual. As we all know from experience, this isn't child's play. What does one worry about? What is one most concerned about with change? Which scenarios develop? Where do you always get stuck? In change management, since nobody is standing securely on the shore, we all have a tendency to flee: uncertainty is kept in check with project plans; personal worries are rationalized; threats are banished by leaping to action. We try to prove that there is no reason to worry. The fact that the intention to change is often hard to understand, that information doesn't flow, and that people simply don't set themselves in motion highlights the power of emotions in change processes. Drawing once more on the iceberg analogy, these emotions sometimes stay beneath the water's surface for a while but always come back with a vengeance. Like a strong gravitational force, they tug the rationality of change management into the depths of the icy sea of emotions.

What do you need to be aware of? The most important rule of thumb with regard to this first category of emotions is probably that negative feelings also need time and space. Before you can cognitively begin dealing with these anxieties, they must first be taken seriously and expressed in a suitable form. You need to find the appropriate language and opportunities to speak frankly with each other.

11.2 ANGER AND AGGRESSION

Flight is not the only option for dealing with threats: you can also stand up to them and fight them. Anger and aggression are proven means for this: people jeering loudly at information events, accusing the team leader of treason at a regular meeting, or cursing "the top brass" in the coffee kitchen. We are here concerned with establishing limits: one's own identity must be staked out, and threats must be held confidently in check.

Anger and aggression are heated feelings that can ignite quickly but can also burn out equally quickly. Like anxiety, when roused, they can stimulate powers you didn't think you had. Anyone who has been confronted by such explosions knows how they get your adrenaline going. Livid e-mails of protest, angry battles of words in meetings, demonstrative shrugging of shoulders, and constant eye rolling—all these have a tendency to make us lose control. This fits in with the classic logic of conflict escalation. Aggression fuels return aggression, particularly in situations where a certain fundamental tension already exists; a small spark is often all that is needed to blow up. This brings to mind a department audit in the banking sector officially concerned with the balance of a year of intensive change. During our interviews with selected employees, quite different topics cropped up:

- **Conflicts** between two departments.
- **Emotional escalations** due to the resignation of an employee.
- **A tendency toward polarization** between fans and opponents of change.
- **Criticism of the management**—apparently the department leader "was never there," had "neglected" the department, and "didn't have any idea of the growing conflicts."

In retrospect, it's now clear to us that the situation needed the audit to release some of the pressure. However, it was the intensity of this release that surprised us. Even during the presentation by two small groups on the subject of "our retrospective," a discussion about motivation struck a chord with the audience. Voices venting anger in familiar words increased, speaking of "frustrating change steps," "chaotic conditions," "demotivation like never before," and "the feeling of being steamrollered by the reorganization." Criticism of the management also became increasingly explicit.

Due to their intensive psychodynamics, anger and aggression are anything but pleasant emotions. As the banking audit amply demonstrates, they work in such a way that they distance you from others. As the department leader concerned put it, it is "as if a storm is directly above you." You try to seek cover and protect yourself.

However, anger and aggression are also important signs that you are on the right track. In other words, their absence would be far more reason to worry. They raise classic change subjects such as the past and the future, collaboration and leadership, and participation and decision.

Anger and aggression also have an important function in radical change processes that one must accept and endure. We must be able to let off steam in order to become more open to new discoveries. In our case study, this was particularly pertinent to messages about trust and collaboration. Subsequently, after completion of the retrospective, we made the decision to continue working on the most important theme clusters in focus groups. The World Café format enabled a high degree of focus, flexibility (everyone is allowed to change table or theme when they want to), and transparency (the most important points of discussion are listed directly on flipcharts on the tables). This not only enables the expression of further negative feelings but also allows change to happen there and then. The personal interaction between the most important partners in this conflict—including the department leader—transformed the problem-oriented energy into one increasingly focused on solutions. Misunderstandings were further clarified, while necessary steps for improvement and different ideas for the future were compared. All of a sudden, people were even laughing together!

HUMOR IN CHANGE MANAGEMENT

Since change processes are full of surprises, they are often met with laughter. This is not only due to the unexpected results change often brings about and the irritations triggered by it but also with funny observations and a good atmosphere.

Our creed is that "change management without humor is pointless." In relaxed situations, anyone can laugh easily. But humor is of particular help when we're concerned with tough aspects of change: serious issues can be discussed less threateningly, important messages communicated more easily, and deadly serious resistance broken down with hearty laughter. Various philosophers observed that "wit" in terms of humor and "wit" in terms of inspiration share the same root. The fundamental interjections in laughter and research or invention are very similar: the "ha-ha" of laughter, the "aha" of comprehension, and the "ah" of discovery are essentially three aspects of the same spectrum.

So what's the moral of this change story? Talking brings people together, which is nowhere seen more clearly than in the context of resistance. When considered from the perspective of the department leader, this case study demonstrates one of the most important leadership tasks in turbulent phases of change: that of coming to terms with your own emotions. Ultimately, the critical factor when dealing with resistance is dealing with yourself. Managers must question themselves; they must endure the tension and at the same time create the right environment to dissolve that tension. Storms have a cleansing effect as long as you ensure that good lightning rods are in place, not least in the form of professional support by experienced change facilitators.

11.3 SADNESS AND DISAPPOINTMENT

Sadness helps you leave certain things behind and move on. In order for something to really come to an end, you must take suitable leave of it. This is neither done by fleeing nor by fighting but rather by beginning to accept at an emotional level that you are currently in a transitional phase. This letting go and saying goodbye is sometimes accompanied by sadness.

Sadness appears above all during personal coaching, where we can work on the acceptance of change in a sheltered environment. This was exactly the case with the former IT department head who lost his leadership role in the course of a large reorganization project. His sadness typically only appeared after a phase of anger: the organization's injustice first had to be violently verbalized, and the new boss cursed to hell. Just as typically, the sadness—a feeling of vulnerability—was triggered by looking back at the past. Like many others who feel they are victims of change, the department head tended to romanticize his former career: "Everything was better earlier" would have been a fitting title for his story. The team was great; the environment motivating; and people valued him for his work: "It was quite simply great fun to come to work every day and carry on working."

Regardless of whether this reflection was really an accurate depiction of the working reality or not, it is important to allow space for sentiments and melancholy in personal change management. Before we can open ourselves up to the future, our present is dominated by memories of different hues that, like anger and aggression, need to be externalized.

In contrast to anger and aggression however, sadness is a slow feeling. Sadness requires time because we need to work through various feelings of loss and departure as well as further cycles of anxiety and anger. Change processes often need precisely this slowing down and repetition before a breakthrough can occur.

Put simply, the more intensive the survival and learning anxieties, the deeper the feelings that should be worked through. Radical changes have a tendency to set the whole iceberg upon us, as in the case of the IT department head we mentioned, who was only able to admit his vulnerability bit by bit. Simply finding the right words required a whole lot of time. How do you describe the feeling of humiliation? Or how do you find a suitable way to express your survival anxiety when you're already 53? How do you vent spleen about the perceived disloyalty of an organization you've been a part of for almost 30 years? Additionally, it takes time to build up trust enough to be open about this sadness and disappointment—sadness and disappointment at the "huge change in the organization," the "status as someone who has lost to the change process," and "all that flexibility that is forever gone." However, the former department head apparently came to the conclusion that this change process was unavoidable. For a real new beginning, you must first get rid of the past; otherwise, you will remain sitting on it like the previously mentioned dead horse. Dismounting from this horse is sometimes associated with social relegation. "For me, it was a gradual liberation," the departmental head told us at the conclusion of our coaching session, "but it was bloody hard work being honest with myself."

Sadness is not only a slow feeling; it is also a muted one. Finding the appropriate language for it helps you free yourself from it. Personal presence and open dialog from management can help cut the cord. Shared rituals such as a farewell party or staged transitions are also established ways of easing sadness.

11.4 ENTHUSIASM, JOY, AND COURAGE

Once negative feelings like fear, anger, and sadness have been adequately processed, there is again space for positive emotions. Real letting go is often followed by a surprisingly large openness to change. One can focus one's energy on learning, practicing the improved processes, and integrating the modified work practices. All strengths can now be brought to bear on new connections, and old, established knowledge appears in a new form. Real reconciliation has been achieved, bridging yesterday, today, and tomorrow.

It is therefore logical that we make as much use of this enthusiasm as possible. Without overdoing the euphoria, advances in the learning process must be made visible and acknowledged. An honest appraisal of these steps can make your change take off as the careful use of success stories demonstrates.

SUCCESS STORIES

Success stories were prominently exhibited in the IT department of an international energy firm. After a workshop on the planned reorganization involving people across the globe, a template that all participants were to use to review their positive change steps was drawn up:

- What is this change step about?
- Who is its sponsor?
- Which stakeholders are involved?
- What has been achieved?
- What's planned next?

The answers had to include images, screenshots, or quotations in order to make the story as comprehensible as possible. Each participant then found an exhibition space for their success story (on the wall, on a column, on the window, and even on the ceiling).

These success stories were exhibited in the form of a public gallery during a department meeting 2 months after the workshop. The exhibition was officially opened by the department manager who had asked for the appraisal. Rather than silently following a typical walk-through, all could follow their own personal curiosity, investigate the stories individually, and discuss them with one another. Within a short time, the room was filled with positive storytelling and feelings of enthusiasm and optimism reigned. "Only now do I know how much positive change can take place!" said one participant, affirming the purpose of this meeting. "And that also changes my perception of improvement steps that didn't work or haven't yet."

Our extensive consultancy experience has taught us that any change initiative not accompanied by emotional mobilization is doomed to failure. Feelings are the driving force, so to speak, of a successful change implementation. Without emotional involvement, without engagement and passion, no change initiative will even get off the ground. This is why it's so important to know the feelings that typically manifest and the setting to which they turn the change motor.

Naturally, we should once more put the brakes on this seductive motor metaphor. Emotions can be as little controlled by pressing a button as organizations are trivial machines. Man as a psychic system and the organization as a social system are too complex for that.

As members of an organization, people are not programmed with a single emotional condition just as the dynamic of change does not follow a single law. Changes are kept on their toes by different emotions at various intensities.

Not all people experience the same feelings at the same time. What leaves one person totally cold might get someone else very worked up. What one person acknowledges with stoic calm causes the second to have doubts and sends the third to the barricades. And while one team internalizes change relatively quickly, another must review everything many times over and switch to and fro between the old and the new for ages.

Apart from the fact that emotions occur at different times with different intensities, we must also be aware of the tendency toward polarization often observed in change processes. While one group of employees might see change as something positive, another group might first and foremost perceive it as a threat. Changes trigger black and white comparisons along the lines of "Everything was better before. In the future, we will have to dress up warm." They can also encourage pressure groups of supporters and opponents.

Emotions appear neither in a united form nor at the same time. We observed this again recently during a kickoff meeting in the health sector. During the opening round when the managers were asked what they were currently focusing on, how they viewed the change process thus far, and what they expected from that meeting, very different affect–logics were shown:

- One manager was **outraged** about what she considered to be a very authoritarian form of change (to quote her, "A dictatorship of change!").

- One colleague felt **threatened by the planned reorganization** "...to the extent that I don't even know whether we'll still be here in 2 months."
- A third manager expressed her **disappointment** "...about the ongoing demise of the previous organizational culture."
- Another department leader described feeling "...**furious about the lack of direction** in the organization." She said it was totally unclear what goals the change was targeting, let alone whether it could provide a convincing vision of the future.

Following this energetic kickoff, very diverse emotions were expressed in the closing round too: satisfaction that "at least an important step towards internal clarity" had occurred, a sliver of confidence that "we just might have found a viable solution for the future," but also pessimism: "We just need to remain realistic. We're not going to gain anything from this change."

As central drivers of organizational processes, emotions themselves are constantly in motion. They can change abruptly and are always good for a surprise or two. For example, unexpectedly positive feelings can appear in the context of change projects as we saw in the case of a financial services provider. Despite skepticism as to the stipulated "improvement of our collaboration" expressed in the initial discussions, in the first coming together of external and internal services, a real sense of get-up-and-go took hold. Collective analysis of the current situation, where criticism and dissatisfaction were allowed to be vented, helped achieve this positive atmosphere. Subsequent comparison of the various visions and their translation into a smart action plan channeled the energy present and put it to good use.

A less pleasant surprise for us was that in the second large group event, this positive energy metamorphosed into its diametrical opposite. The presentation of the intermediate results received as little recognition as the lessons learned from the measures that had already been implemented. What conclusion did we draw from this? Things do not always happen as expected. In other words, beware of confusing an atmosphere of get-up-and-go with successful change. The case study reminds us that with change processes we're often presented with a hot-and-cold bath. Whether we see this bath as an external consultant, an internal change manager, a line manager, or an employee, dealing with swinging emotions in a professional manner remains a great challenge.

The unpredictability of change processes highlights the importance of an agile approach in change management. What are the significant warning symptoms indicating necessary changes? What does our client demand? In which areas do we urgently need to improve? But also: Who is currently occupied with what? What is the currently dominant emotional state? And what do we need to watch out for in particular, in order to focus on the right things without losing anyone in the change process?

Behavioral routines that determine the current business can be seen as a type of emotional freeway from the systemic perspective. Whatever's always been done a particular way is not only associated with behavioral patterns but also with deep-rooted emotional patterns. On the one hand, all people drive in their own lanes, while on the other hand, it's rarely the case that drivers on the freeway move independently of each other. This means that specific routes inevitably evolve. All this indicates the enormous effort required to build a new freeway as part of a change process. It is very likely that this new construction will cause a variety of traffic disruptions: blocked lanes, diversions, queues, waste of time, increased chance of accidents, and so on and so forth.

Professional change management demands that the correct signs are put in place to divert us from individual movements on the freeways of emotions to what one could call the traffic regulations of an organization—in other words organizational culture.

WHAT YOU CAN TAKE AWAY FROM THIS CHAPTER

Organizational change is inseparable from personal emotions, which can be seen as the elixir of any change: they energize; they give meaning; they drive.

We are usually concerned with four categories of feelings in change processes:

1. Uncertainty, worry, and anxiety
2. Anger and aggression
3. Sadness and disappointment
4. Enthusiasm, joy, and courage

It is mandatory to understand these feelings correctly in order to get change up and running. Firstly, a profound recognition of the various forms, dynamics, and functions of these emotions constitutes the basis of successful change management.

However, beware of illusions of control. Emotions appear neither in the same form nor at the same time. They need different amounts of time, space, and attention. They are unpredictable and there is no magic formula that eliminates all emotional challenges.

Nevertheless, change management cannot occur without emotions, because without emotional mobilization, no change initiative will make any headway at all.

12

CORPORATE CULTURE AND POLITICS

What does all we've been talking about have to do with culture, then? And what type of culture are we talking about when we discuss corporate culture? As is the case with the definition of the concept "organization," we immediately seem to know precisely what we mean. Our everyday understanding presents us with all sorts of artifacts that generate the culture of an organization, for example, the brand name, the architecture of a company's offices, the workspaces, the nature of the advertising, or the experience of a particular product. In fact, all these elements play a part in the definition of culture. The model inspired by leadership guru Edgar H. Schein [1] makes it clear that corporate culture is much more than a loose combination of individual elements.

Figure 12.1 demonstrates that culture:

- Is influenced by the **context** of an organizational system. Corporate culture isn't a monument set in stone. It reacts much more selectively to each market situation, the political climate, the competition, or social change.
- Is driven by a particular **mission**, a specific organizational purpose, and is oriented toward a particular **vision**.
- Exists in the **triangle formed by goals, structures, and behavior**.
- Can be subdivided into three levels:
    1. **Artifacts**, that is, all phenomena that you can see, hear, or sense, such as the architecture, décor, products, and style of an organization. The artifact level also encompasses all processes that make this behavior routine.

Kanban Change Leadership: Creating a Culture of Continuous Improvement,
First Edition. Klaus Leopold and Siegfried Kaltenecker.

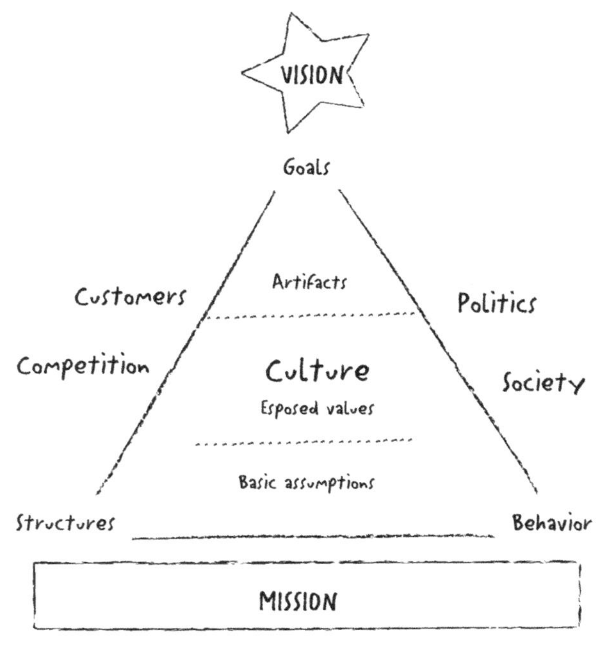

FIGURE 12.1 Corporate culture.

2. **Espoused values**, that is, all messages that are essentially guidelines: the organization's goals, the official strategies for achieving these goals, but also the values or philosophies presented to the outside world ("We stand for...," "We represent...," "We handle...," etc.).
3. **Basic assumptions** that, as largely unconscious, self-evident stipulations, form the parameters for actual behavior: What does the organization's management really expect? What behavior do my colleagues particularly want to see? How do people in this team communicate? What is frowned upon here?

The culture determines the attention given to something in terms of the market challenges or the improvements to organizational collaboration.

If one of my basic assumptions is that I alone as CEO determine the orientation of the organization, my conclusions will be different to those resulting from several perspectives determining the orientation. If I am convinced that individual performance is ultimately all that matters, teamwork would have a different value than if I were to treat it as the lifeblood of my corporate activity. And if I have always been programmed to value exclusive market leadership through internal research and development, then the approach of open innovation won't seem so attractive.

How does this cultural model help us shape change professionally? It helps by making us aware of a classic trap, namely, that of treating internal PR as the be-all and end-all of communication. If the CIO of a media organization is nonplussed by the fact that he has already explained the new strategic orientation three times and despite this nobody seems to pay any heed, then he has fallen into this trap. The same is true of the new department head who laments the fact that a certain team's collaboration has not improved an iota despite her repeated appeals to the members' team spirit. And again of the experienced change manager who fails to see his sophisticated project plans score in strengthening marketing and sales. Because, as Ed Schein's model clearly shows, it is the basic assumptions that determine how things really happen.

The fact that culture is at the heart of the corporate model underlines its significance. Culture encompasses all values and convictions that define our corporate actions. It is everything we believe to be meaningful. Culture bundles together mutual expectations and brings together all of those obvious truths that need neither be spoken out loud nor explained: "That's just how things are here!"

FIGURE 12.2 What is culture?

Ed Schein uses a wonderful metaphor here: culture is what makes us feel at home, just like a fish in water. Hence, as Figure 12.2 shows, we're pretty surprised when we're asked to describe this completely natural state of affairs. Three points can be drawn from this:

1. **Culture is deep**. If we only observe the surface and believe we can simply manipulate the culture as we see fit, then failure is guaranteed from the outset. Moreover, culture controls an organization more strongly than the organization can control culture. And it's a good thing too, because ultimately it's culture that gives our daily business stability. As members of an organization, we learn what works and we develop convictions that gradually enter our subconscious. As basic assumptions about the world in general and work in particular, they regulate how things are perceived and approached in the particular organization.
2. **Culture is broad**. An organization learns to survive in its environment, and its members learn to survive in their organization. Certain rules of play are

necessary in both cases, and in the course of time, these rules will come to mimic laws of nature: that's how you deal with the boss here; that's how you deal with your colleagues and customers; that's what you need to do to advance your career; that's how you get the sack; those are the holy cows; those are the skeletons in the closet; etc.

3. **Culture is stable**. The members of a group want to hold onto their basic assumptions because culture creates meaning and makes life predictable. Most people don't want unstable situations and go to great lengths to attain the greatest possible level of consistency.

12.1 THE POWER OF CORPORATE CULTURE

Ed Schein's model has far-reaching consequences for anyone intending to create a culture of continuous improvement. It explains why a corporate culture can't be changed all that easily. After all, it represents the collective knowledge of a group of people, in other words the thoughts, feelings, and perceptions that have made a group successful. This model also makes clear to us that significant portions of the culture are invisible. At this deeper level, culture can be understood as a shared mental model that the employees of an organization represent. In effect, they cannot describe their culture—just like the fish in Figure 12.2 who, even if they could speak, wouldn't be able to explain what water is.

MENTAL MODELS

In his five disciplines of a learning organization, the well-known MIT professor Peter Senge also accorded culture central importance [2]. For Senge, corporate culture is the sum of all mental models that have a large amount of influence as deeply rooted assumptions of how the world is perceived in a certain organization and how people act accordingly. As basic elements of any corporate culture, mental models are therefore:

- Like a sort of glass window that frames our perceptions and subtly distorts them
- Pointers in a complex world
- Maps that make a certain amount of navigation possible
- Assumptions we have shaped into convictions
- By definition full of error and incomplete
- Normally not checked or scrutinized

Mental models are adequately described by the joke about the driver who hears a warning on the radio about someone driving on the wrong side of the road and then roars, "What d'ya mean 'someone'? There're hundreds of 'em!"

Although corporate culture is remarkable for its high level of stability, it is by no means impossible to change. Quite the opposite, the culture of an organization is subject to a specific life cycle. Initially, every organization establishes a pioneer or start-up culture determined by the values of the founders. If the organization grows and if it achieves success in the market, then its culture will also change. Alignments will have to be made; values will need to be redefined; subcultures will evolve. Aging organizations ultimately tend to perceive relationships that were created at some point in time as ironclad principles. Culture becomes cemented, to an extent, and begins to hinder change. "We've always done this like this!" "That won't work here." "That doesn't make any sense for our organization." Such statements are frequently heard—evidence of these blockers of change.

But what's the real meaning of culture, as far as continuous improvement is concerned? The meaning of culture, one is preprogrammed to answer, is that it is a central factor for the success of an organization. It is not some sort of "nice-to-have" phenomenon, something that is to be celebrated only on holidays and be observed only by niche departments.

It is much more the case that the culture determines how we deal with the external and internal demands on an organization. In doing so, a culture also determines how change is to be managed and how kaizen can be achieved.

Seen from a cultural viewpoint, it is necessary to come to terms with the paradox that is organizational development. The more an organization learns how it can successfully master the relevant demands and stabilize certain patterns of success, the harder it will be to question those patterns. Until, that is, the organization has unlearned how to set itself new demands, orientate itself better strategically, hone its structures, and establish new relational forms—in short, to change itself.

Seen from a systemic viewpoint, we are less concerned with what constitutes a culture. More important is the question whether it is functional, that is, it promotes the strategies, processes, relational forms, and behavior necessary to successfully surmount the current challenges. Whether the culture promotes or hinders the business success of an organization depends on how consistent its artifacts, espoused values, and basic assumptions are. If these three levels do not fit together optimally, then the cultural pulse is weakened.

Actually, in many organizations, the cultural pulse is plagued by all kinds of disruptions. This happens mainly because contradictions between artifacts and espoused values are overlooked. For example:

- People **talk constantly about openness** although the employees sit in small cubicles with their **doors permanently shut**.
- The **importance of communication** is emphasized at all official occasions although informal exchange is labeled as **mere chitchat**.

- **Collaboration** is heralded as a key value, but **nothing is done** to establish it at a silo or hierarchy-bridging level.

Similar disruptions to the cultural pulse are also evident between espoused values and concrete behavior. Organizations:

- **Praise the virtues of teamwork** although they base all rewards and control systems on individual responsibility. This suggests that **when it comes to the crunch, it is the individual** and not the team that counts.
- Commence **organization-wide diversity management training programs** while ignoring the **income gap** between men and women, the **career disadvantages** for migrants, or the **rejection** of older employees. This again leads one to conclude that the work of young, white, native men is valued higher, as has always been the case.
- Officially appear to value **the perspectives of employees** but reduce their participation to **anonymous written surveys**, which says rather a lot about organizational trust.

Even when almost all principles such as teamwork, open communication, empowered employees, or consensus-oriented decision-making are propagated, it is still a fact that these principles are not practiced in many corporate cultures. Rather, principles of hierarchy and command and control are pursued, as usual.

Although corporate culture for the most part is based on underlying assumptions, it is neither inaccessible nor unchangeable. If a change initiative is to be successful, it must afford culture sufficient attention. One must ask about the extent to which assumptions strengthen or hinder the values of kaizen. If an organization wants to establish a culture of continuous improvement, a leadership culture is absolutely necessary. As we will see in further detail, people must communicate and make decisions that are comprehensible in terms of what should be improved, by whom, when, and how.

KAIZEN

The concept of kaizen comes from the Japanese and literally means "change for the better." Kaizen means far more than just technical change management; it is a philosophy for life and work at the center of which lies the quest for constant improvement. Orientating all organizational processes toward customer value is thus just as critical as cultivating the right attitude. Process optimization goes hand in hand with a humanization fed by professional leadership, which always begins with the individual and never ends. Leadership values such as identifying opportunity, positive thinking, self-responsibility, and a solution orientation are an ever-present challenge.

12.2 CORPORATE CULTURE AND MICROPOLITICS

If number-obsessed control, one-way communication, and deficit orientation are part of the daily agenda, then improvement initiatives such as Kanban are doomed to failure even if they are officially backed up. Perhaps visualizations, WiP limits, or service classes will be implemented with some effort, but as long as the cultural laws remain untouched, evolutionary change management won't produce the desired results.

We know very well from daily organizational life that desired results are often not achieved due to organizational politics or rather the ruling cultural dynamic usually termed "micropolitics" or "power games."

Why is this? On the one hand, we can explain it with the fact that neither the management nor the employees are selfless agents of the organization. One tries to create possibilities to exercise influence, to secure these possibilities, and if possible to increase their number. On the other hand, this phenomenon can be explained with the fact that, as economist Oswald Neuberger writes, in change processes there are always several participants who have differing perspectives, information, interests, values, and goals. Each way of acting is fundamentally contingent and could thus also differ and must therefore be justified and defended.

The myriad interests emphasize the fact that in change processes it is not only the *need*, *ability*, and *willingness* that are important; the *permission* is also a central criterion.

Is the desired change allowed? Are all relevant stakeholders in agreement? Is professional change management being promoted? Does the initiative enjoy the attention and active support of those in power? Or is it more the proposal of a group on the periphery? Are all decision makers pulling in the same direction? Or is the change initiative a bone of contention that could easily leave one caught in the middle?

"Power games" don't refer to merely passing the time in a leisurely manner ("play") but rather to a ritualized contest ("game")—a contest that:

- Is **unjust**, meaning that particular players, by the very nature of the rules of play, have a lower chance of winning
- Is **asymmetric**, meaning that swapping the players around would change the game
- **Remains indeterminate**, meaning that, due to incomplete information, several solutions are always admissible
- Is **context and person dependent**
- **Is based on illusion and bluffing**, for example, via suppressing certain information through manipulation, sparking enthusiasm by ignoring critical concerns, stockpiling, hidden foul play, and much more [3]

In this respect, corporate politics is a perfect example of the latent self-contradiction of corporate culture. On the one hand, there is the officially extolled, often highly staged, macropolitical commitment to this, that, and the other. On the other hand, there are micropolitical dynamics, characterized by a lack of transparency. This is because in the case of micropolitics there exists a "Rumpelstiltskin effect: the moment it is referred to by its name (i.e., exposed) it loses much of its power" [4].

We are therefore here concerned with the consistent excommunication of the topic. We mustn't talk about power, the political game mustn't be named, and its powers must not be pondered. As a plan, it is the opposite of German politician Walter Fisch's catchphrase, "Do good deeds and talk about them," and reads, "Act politically and keep silent." A lack of transparency, obstruction as an inherent necessity, and obstinate denial all make micropolitics very hard to observe: it is as difficult to identify in official organizational artifacts as in formal daily behavior. In order to identify it, you must look behind the façade. You must to a certain extent—apropos basic or underlying assumptions—go underground.

Nevertheless, political action in organizations is not at all a different means for continuing the struggle. Quite the opposite: acting politically is oriented around a structured, collaborative setup. According to Neuberger, micropolitics is no workplace accident but a constitutive part of organizations:

> Political action does not mean chaos, born of a struggle of each against the other in self-hungry pursuit of interests. Rather the opposite: political action is oriented around a social ordering. But not a mechanical or bureaucratic one; rather, one that results from the "play" of powers. Such an ordering is flexible; it adapts itself to new situations more quickly than a rigidly formal system of rules [4, p. 714].

Obviously, if everyone were to stick strictly to the rules, nothing would work in organizations at all. "Work to rule" is still one of the greatest hindrances to organizational success. Success can only be expected if someone takes responsibility for a task and implements it with a wealth of ideas and the joy of getting stuck in, even if this means that the person concerned needs to exceed work-to-rule stipulations, overcome treacherous terrain, or discover new territory.

The deciding factor for such implementations is the impact of micropolitics. Whether power games are functional or dysfunctional is of great significance in change processes, just as for organizational culture as a whole. Do they help enforce the right powers or do they hinder them? Is improvement encouraged or disrupted? Do the current politics allow the creation of a kaizen culture or do influential stakeholders feel themselves threatened because of it? Does the management feel called into question? Do the opinion leaders feel negated?

In the following chapters, we will answer these questions and in doing so delve even deeper into the world of Kanban.

WHAT YOU CAN TAKE AWAY FROM THIS CHAPTER

Corporate culture is far more than a loose collection of cultural artifacts. It unites the cornerstones of corporate mission, organizational structures, and personal conduct. Culture encompasses every aspect of how things work in an organization. Similar to the iceberg model, organizational culture has three levels: artifacts, espoused values, and basic assumptions. While the first two levels are visible and audible, the basic assumptions remain mostly unspoken and often unconscious too.

- What do people in this organization believe?
- What are the underlying assumptions?
- How do people approach certain things?
- What are the character-defining features of the organization?

The answers to these questions give direction to our behavior.

Corporate culture is a phenomenon that is as deep as it is stable. It is like a motorway on which ruts have appeared through the course of time, keeping our perception, thoughts, and actions on particular routes.

The deciding factor for the success of change projects is whether these cultural routes are functional or not. Do they help process the desired change productively? Or do they rather block the change? Do we first need to build new motorways in order to enable continuous improvement?

Many improvement initiatives fail because the power of corporate culture is not acknowledged at all, or at least not acknowledged enough. Ignoring micropolitical behavior and the dysfunctionality of power games in change processes contributes to this problem.

13

CONCLUSIONS FOR KANBAN CHANGE LEADERSHIP

What are the conclusions from all we've seen so far for Kanban change leadership? Which forms of leadership are necessary for the creation of a culture of continuous improvement? Given the complexity of today's business world, if we do not want to give up at the first stumbling block, there are three important conclusions that can be drawn from what we've seen so far. In today's change management, we are primarily concerned with:

1. **Mindfulness** of what is happening in the market and in the organization, that is, observing, hearing, having a good feel for things, and meeting customers' desires
2. **Professional communication**, outward as well as inward, with all relevant stakeholders across organizational boundaries, hierarchical levels, technical departments, and beyond
3. **Agile modeling of the change process** through the application of state-of-the-art methods

Mindfulness, professional communication, and agile modeling of processes are the three central factors of any promising Kanban change management, as well as being our survival kit for the cultural journey toward continuous improvement. But this needs a little explanation.

Kanban Change Leadership: Creating a Culture of Continuous Improvement,
First Edition. Klaus Leopold and Siegfried Kaltenecker.

13.1 MINDFULNESS

As we have already extensively shown, Kanban begins with mindfulness. A good overview of the current situation is at the heart of this mindfulness. Current work processes are examined from various perspectives and made visible on the Kanban board. We are here concerned with a well-structured representation of the value chain that considers the key factors for all significant partners.

Consequently, in every area of work in which Kanban is introduced, the process is initiated with an eye on the entire system. We are concerned with an organization-wide focus on the current situation at various levels.

- With a **focus on the market and the internal organization of processes**: How smoothly do our value chain processes run? Which current opportunities and problem areas can we identify? How well are we processing them?
- With a **focus on the organizational setup**: How well equipped are we to fulfill our tasks? How functional are our structures? How clear are our responsibilities?
- With a **focus on real collaboration**: How effectively do we cooperate? How suitable is our culture? What significance does improvement have for us?
- With a **focus on individuals** who keep an eye on their own reactions: What is changing for me? What do I perceive? How do I feel about this? And what conclusions can I draw from it?
- Last but not least, with a **focus on one's own understanding of leadership**.

The Viennese organizational theorist Rudi Wimmer defined management's willingness to begin this perception with themselves as a key factor for every type of change management [1]. The first thing necessary for the creation of a basis for successful change is management understanding that it cannot propagate itself as a change agent without simultaneously considering necessary change in its own leadership performance. According to Wimmer, an adequate perception of current market opportunities and risks is not the only thing that is important here: the existing corporate culture must also be examined and any possible dysfunctionality in management detected.

Based on our own observations, however, this is anything but obvious. Apparently, few managers begin this change with themselves. It seems to be more the case that other people are primarily challenged, made to feel uncertain, and called into question. Need an example?

According to a recent survey of 1100 British first-line managers, a full 72% of those questioned never have any doubts about their own leadership capabilities.

This symptom can be explained both by a management-specific tendency to overestimate its own capabilities and by the fact that 80% of all managers would put themselves in the best 20% [2]. But what lies behind such distorted self-images? What observational pattern is at play here? What cultural assumptions are such perceptions based on? It wouldn't appear too far removed for us to discover the archetype of the genius conductor, the confident sea captain, or the all-knowing general behind this. The burden of a mechanistic image of the organization seems quite obviously to correspond to the inheritance of a leadership model that exerts itself neither in the field of critical self-reflection nor in that of suitable humility. On top of this, there is also the mental model of management French management pioneer Henri Fayol established almost 100 years ago: plan, organize, command, coordinate, and control [3]. To put it more colorfully, if you think of an organization as a machine, then the manager is a mechanic and administrator.

In the course of the twentieth century, the demands on organizations changed just as radically as the demands on their management teams. Despite the radical changes in the fields of technology, lifestyle, and geopolitics, the dominant leadership paradigm has more or less remained the same. According to the renowned management thinker Gary Hamel, leadership is based on the mental model of rational *business administration*, just like it always was [4]. Goal orientation, process coordination, and result control remain important objectives in the twenty-first century. However, the hierarchical bureaucratic model of management has in many ways become a little aged.

What characterizes good management today? "You are supposed to figure it out for yourself, like sex," said Henry Mintzberg [5]. Even though he doesn't give us a recipe for success, Mintzberg at least reveals a few ingredients.

Fundamentally, effective management is a combination of art, science, and craft.

Figure 13.1 outlines these cornerstones, distilled by Mintzberg from his own observations of the daily business life of various managers, characterized by speed, fragmentation, and discontinuity. The various communication demands mercilessly wash the image of a confident sea captain overboard. At this point, there is even talk of "the end of management," completely obsolete technologies, or counterproductive routines. In the future, according to current opinions, leadership won't be focused on administration but rather on continuous improvement and innovation.

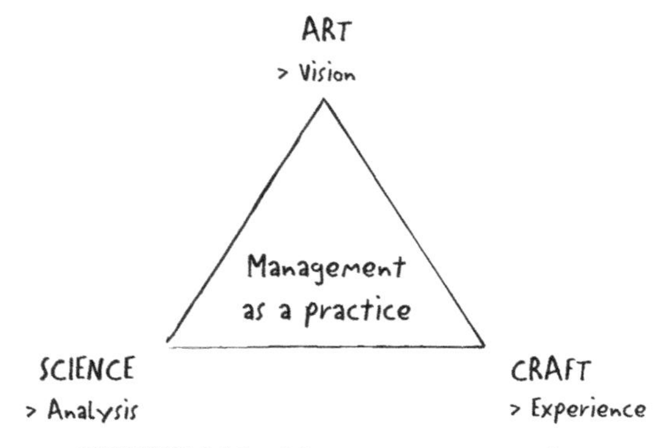

FIGURE 13.1 Management as a practice.

TABLE 13.1 Models of Leadership

Twentieth century	Twenty-first century
Primarily orientated around shareholder value	Balanced orientation around all stakeholders
Administration orientated toward short-term profit	Orientation toward long-term guarantee of success through continuous improvement and innovation
Command and control	Observation, communication, facilitation
Hierarchy and bureaucracy	Lean networks
One-way communication through directives and reporting	Two-way communication through reflection and dialog
Authority as position	Authority as practice
Centralized leadership	Polycentric leadership
Individual decisions: top-down	Leadership as a team sport: cross-functional and hierarchy-bridging decisions
Manager as hero and conductor	Manager as enabler and coach

13.1.1 A New Paradigm for Management and Leadership

The comparison between the dominant models of leadership in the twentieth and twenty-first centuries in Table 13.1 highlights their key differences. The exclusive orientation around shareholder value gives way to a balanced orientation around the interests of all stakeholders; a focus on short-term profit morphs into a long-term perspective; the primacy of rational administration yields to the aspiration to create a culture of continuous improvement and innovation.

Thus, the former administrator of standardized business processes becomes an organizational designer for high-performing teams [6]. The ability to set clear goals, establish modes of decision-making, and free up resources is also a part of this, as well as the ability to develop a constructive way of dealing with your own limitations and ignorance. In this context, professionalism means using the inevitable uncertainties, confusion, and cluelessness as resources.

In the twenty-first century, command and control gives way to a culture that respects self-control without losing sight of the organization-wide need for coordination. This is the quintessence of the contemporary management and leadership debate.

New forms of network-oriented leadership appear alongside hierarchical management in order to use the available expertise optimally, particularly with respect to the current perception of environmental dynamics. "Leadership as a team sport" becomes a key factor.

We confirmed that leadership is a team sport during our study "Successful leadership in an agile world," where we questioned 58 practitioners of agile software development about their experiences [7]. As Figure 13.2 shows, the concept of team sport leadership mostly indicates two necessary changes: both the centralism of traditional leadership and the one-sidedness of communication and decision routes must be overcome.

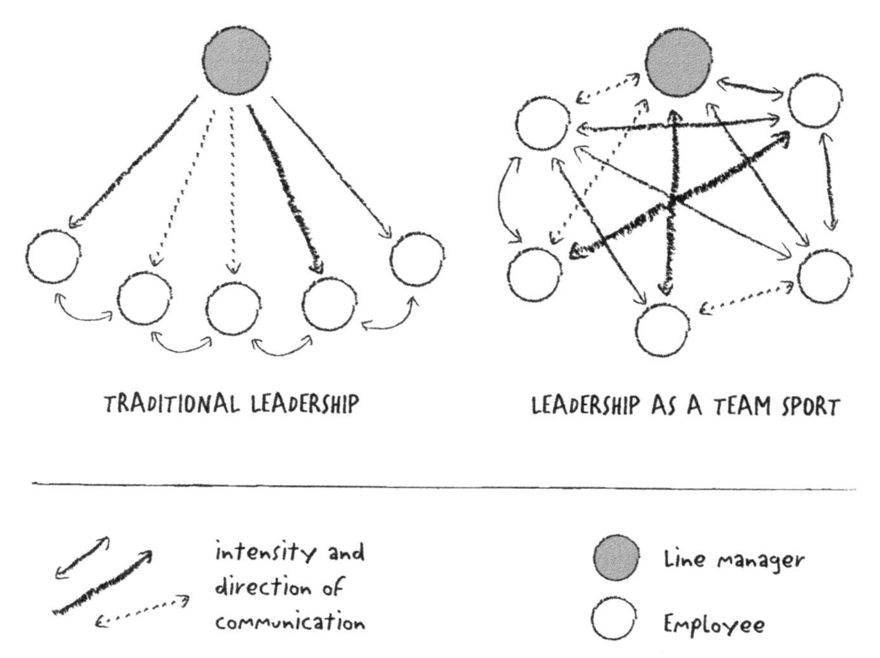

FIGURE 13.2 Traditional leadership and leadership as a team sport.

In soccer terms, a coach backs a team that, by working for each other, achieves victory. Thus, the dictatorial "Pass the ball to me!" of the supposed captain is replaced by flexible play, on and off the ball. At the end of the day, a defender can also score goals, and an attacker can also deflect a dangerous situation in the team's own penalty area.

Let's return from the soccer pitch to the world of the organization: In the context of the organization, team-oriented leadership has far-reaching consequences. It demonstrates that the success of an organization is a result of the flexible cooperation of highly varied forces, whose effectiveness doesn't depend on the formal position ("I alone am the management") but rather on the relevant competences in the situation at hand ("In this game situation, I provide leadership"). Figure 13.2 highlights how, beyond hierarchical orderings, a variety of formal and informal leadership processes are deployed.

The reports from the practitioners interviewed for our study resonate with our own experience. As managers and consultants, we have in various situations observed that the close collaboration of different technical and leadership experts is a decisive factor for top results. Furthermore, feedback from these practitioners resonates with theoretical discussions of leadership as a trait of the system rather than that of an individual manager. With the publication of Katzenbach and Smith's classic, *The Wisdom of Teams: Creating the High-Performing Organization* [8], shared leadership became a part of the lexicon. Shared leadership means that all team members:

- Take on **responsibility for the overall success** just as much as for individual development

- Achieve and "sell" **results together**
- Distribute authority situationally in favor of technical competence
- Establish **network-like communication** models
- Bring decisions approved by all colleagues to the center of actions
- Force **critical examination of work processes** and, if necessary, adaptation
- Subject the **quality of collaboration** to regular consideration

With his model of a "leaderful practice," Joseph A. Raelin goes a step further. He defines four qualities of contemporary leadership: it is *concurrent* in terms of the simultaneity of leadership performance, *collective* in terms of a shared responsibility that cannot be delegated to disciplinary superiors, *collaborative* in terms of intensive teamwork, and *compassionate* in terms of each team member supporting the other. According to Raelin, we need organizations that give everyone the possibility to become leaders based on situational demands and individual expertise [9].

But what is this to Kanban change management? What practical pointers can we extract from the contemporary leadership debate? What we need is a new, cultivated form of mindfulness that encompasses as many areas as possible:

1. **All involved in a particular creation of value**. Kanban change management must consider all relevant stakeholders right from the start and keep an eye on them because customers, managers, employees, and suppliers are all living parts of a culture of continuous improvement.
2. **Bottlenecks in processes**. If we understand Kanban as a sociotechnical system, these bottlenecks can appear in completely disparate areas: in the area of unsuitable WiP limits or service classes, as well as in the area of personal attitudes or leadership performance.
3. **The workflow**. Kanban change management tries to disrupt the workflow as little as possible by implementing small improvement steps, aiming for a good flow of change.
4. **Goal-oriented collaboration**. For many change managers who work with Kanban, "leadership as a team sport" is a guiding metaphor. Firstly, it makes clear that we're not just concerned with the core team here but rather with all relevant players in a particular value chain. Secondly, it highlights the fact that each player contributes acts of leadership.

All told, we are convinced that Kanban should be introduced with a contemporary understanding of management. This is as true for the preparation of the first change steps as it is for sustainable operations.

This means that leadership learning—the willingness to put your own patterns of communication and conflict up for regular inspection—constitutes a core value of kaizen practice.

13.2 COMMUNICATION

If effective collaboration is the quintessence of leadership, then change management is concerned with more than just individual perceptions. These perceptions must be collectivized such that they transcend both expertise and hierarchy. According to Rudi Wimmer, the management culture of an organization is key:

> Leadership structures, the interplay of the management on and between the individual hierarchic levels, their level of qualification—all these aspects are an essential part of the problem that is to be processed through the transformation of an organization. It is precisely this point that turns management's ability to change, its willingness to start with itself, into a decisive bottleneck for deep change initiatives. Without notable change in the nature of the leadership, organizational changes aren't going to happen [10].

But how do we achieve such changes? What should you begin with, if you want to begin with yourself? And what should be communicated, by whom, and in what form? Just like perception, communication is also bidirectional. Further guidance is provided by the three-faced, change-critical monster: "Do we need to do it?" "Can we do it?" and "Do we want to do it?" Following is an overview of the corresponding work program [11]:

Do we need to change?

- Representing the **need for change**: what could happen to us if we carry on as we've been?
- **Affect people**: giving people of a sense of urgency.
- Elucidate **positive aspects** of the change.
- Discuss **critical factors** in the change process: What are the risks in making the change?
- Creating a **balance between safety and uncertainty**: showing what needs to be modeled anew and what can be used for further development.
- Representing **planned, step-by-step sequences**: What are we aiming for? What do we need to be aware of?

Can we change?

- Offer **concrete assistance** for the learning process: What events are we planning? What training and coaching can we provide? How are we supporting the learning that is needed?
- Use **key players and multipliers** to bring together a strong change team and support it.
- Give **regular feedback** about the improvement steps: How do we measure and assess these steps? What joint retrospectives do we have?
- **Assist with uncertainties** and provide encouragement: Can we do that? Yes we can!
- **Be aware of your own uncertainties** and allow enough time to process them.

Do we want to change?

- **Keep role modeling in mind**: water doesn't flow uphill!
- Clarify **attitudes**: What is our mental collaboration model like? What does it say about the culture of our organization? What are our underlying assumptions with regard to change?
- Conduct **personal dialogs**: open and frank discussions about hopes, concerns, ideas and questions for the future, concrete impulses, and crazy visions.
- Focus on the **benefits and opportunities**: by focusing on something, we make it stronger!
- *Professional change management is fostered by communication with all stakeholders.*

This raises a few questions that aren't always easy to answer: Who are our stakeholders, actually? What drives our partners in the value chain? What interests are they pursuing? And how are these interests related to each other? All these questions are critical for Kanban, too. If they are processed seriously and not simply reduced to on-the-fly discussions, call-center questionnaires, or even e-mail surveys, then the existing interests can be identified as early as possible. In today's world, organizations require the right kind of agility, first and foremost an accurate responsiveness to ever-changing markets and customer expectations. This responsiveness ultimately boils down to permanent strategic scanning, having your feelers everywhere, and remaining attentive. These are the factors that decide between corporate success and failure nowadays.

Kanban change leadership strengthens responsiveness by having the stakeholders participate in shaping the change initiative right from the start. In Part 3 of this book, we extensively explain how to identify and get in touch with them, how to campaign for the change, and how to achieve lasting agreements. At present, we will content ourselves with a pointer to the systemic spirit of continuous improvement. This spirit is summoned with the first Kanban impulse and continuously drives the evolutionary change process. This spirit manifests itself in the form of visualizations, WiP limits, and service classes; it is nurtured through regular meetings, insightful dialogs, and various learning loops. Kaizen is thus attained and strategic agility reinforced.

Kathleen Sutcliffe and Karl Weick have appropriately labeled this "managing the unexpected" [12]. Figure 13.3 provides an overview of how to achieve high performance in an age of complexity. Whether it's driven by the strategy of acting with anticipation or by the strategy of containing the unexpected when it occurs, the **culture of mindfulness** Sutcliffe and Weick outline in their fascinating book is certainly not only a building block of the so-called high-reliability organizations such as aircraft carriers, power plants, or emergency rooms. Preoccupation with failure, reluctance to simplify interpretations, sensitivity to operations, commitment to resilience, and deference to expertise are indeed guidelines for any culture of improvement.

In practice, however, there is scarcely an organization that is constructed such that unexpected events can be well managed. Business as usual consists neither of a

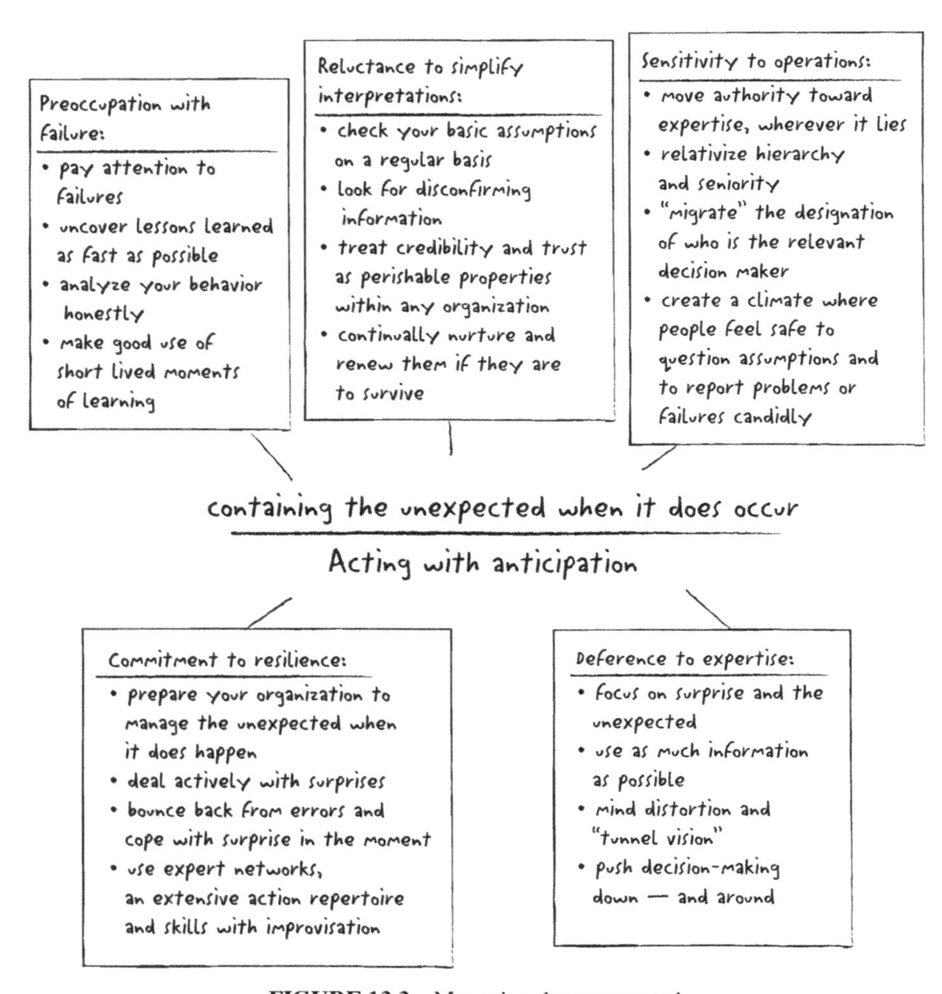

FIGURE 13.3 Managing the unexpected.

systematic learning from failure nor an expertise-led delegation of authority. Agility is of course praised the world over, but it's rarely achieved in standardized processes that are not situationally required, let alone continuously improved. Once again, this has more to do with organizations' systemic ignorance that represses disquieting external impulses as easily as it does the need for critical introspection. Many organizations instead rely on tried and tested structures and more or less avoid open and frank communication. "Stakeholder orientation" is often nothing more than a word; customer satisfaction or even "customer delight" [13] merely a nice idea. Corporate reality far more resembles the old office saw: "The only one causing trouble here is the customer."

Behind all this lurks the fear of conflicts that cannot be controlled: fear of the unexpected. If command and control is taken as a cornerstone of corporate management, then communication must be limited to planned routes, hence controlled. Consider the case of the IT department in the public service. The department manager and the four team leaders in this department were a prime example of the antithesis of leadership as a team sport. Each had more or less only their own interests at heart; a high level of mistrust reigned; communication only happened sporadically; and the group meetings were evidently primarily information events. Fittingly, they also made sure that nothing unplanned happened at the start of the planned departmental workshop. After two official speeches, a series of inputs was devised according to which the respective teams had to process predefined questions. The answers were presented staccato, swiftly, one after the other. An open discussion had not been planned. The management's discomfort was palpable when one group raised critical questions on the future of their department during the presentation. Two team leaders felt obliged to throw the group a couple of quick-fix answers: critical points were deflected and a more in-depth discussion was declared superfluous. "That's on the agenda for our next management meeting anyway," said one of the team leaders finally. "We shall of course let you know what our conclusions are!"

The course of this event—officially called a strategy retreat—clearly demonstrates that open communication can set off deep-seated fears. But whoever asks serious questions risks receiving answers he or she perhaps doesn't really want to hear.

The retreat described here is unfortunately not an exception. In many places, critical objections are staved off. Diverging opinions, the appearance of dissonance, and even the manifestation of conflicts are themselves for many managers shocking experiences accompanied by feelings of loss of competence. That these results are not used as resources but rather avoided highlights the emotional dynamic of these changes.

The emotional dynamic highlights the fact that alongside observation and listening skills the communication and conflict capabilities represent a decisive bottleneck for any change effort.

"The greatest enemy of communication is the illusion of it," Paul Watzlawick once told us [14]. If we run through our own experience of communication, we can only agree. We shudder when we remember a workshop with the top managers of an organization that specialized in online entertainment. Officially, we were there to strengthen teamwork and bring agreement on real steps for improvement. In actual fact, however, there was less talking done with each other and far more with no one in particular. We were occupied for endless hours with an agreement concerning a binding interface management that was stuck due to an excess of special interests. Unsurprisingly, things got pretty heated in this micropolitical arena: everyone interrupted; few listened to anything. Our attempts at facilitation, in order to find an exit from this arena and move from a discussion at the content level to one at the relationship level, failed. Communication became a complete illusion.

So the question is: How can you avoid this illusion? Or to put it more positively: How can real communication be encouraged? And how can Kanban be of help here? Kanban first of all tries to secure this communication structurally: the daily stand-up, queue replenishment, retrospective, and operations review all belong to the standard program of continuous improvement. On top of all this is the fact that even in the first phase of your Kanban introduction, you should be communicating intensively with your team, with your bosses, with your value-adding partners, with peers from other areas of work, as well as with external coaches. Kanban change management needs all this interaction. It doesn't just follow the exclusive plans of a process engineer but rather the ideas of all stakeholders involved. Ultimately, everyone involved in the development of highly complex software these days is highly intelligent and self-confident. Neither does the team, nor the customer, nor the management need someone to do the thinking for them. It is far more the case that the "what" and the "how" of development are determined by those involved in the change process. In fact, one could claim:

Mindful communication is the first bottleneck that a Kanban initiative must master.

That communication takes place doesn't yet say anything about the quality of that communication. So what type of communication should Kanban change agents aim for? How should we behave when making contact with other stakeholders? And in what way should communication be improvement oriented? Our experience is that it is the culture of reflection and dialog that provides useful answers to these questions. Allow us another short sojourn into the realm of systemic thinking in order to get to be bottom of this culture.

13.2.1 The Meaning of Joint Reflection

One of the core principles of systems theory is that reflection is the heart of organizational self-control. This doesn't necessarily mean that this reflection must take place consciously. Often enough, it doesn't even play an important role in change management. Necessary changes to the organization nevertheless take place constantly, but the question is in what way and with what results. Conscious reflection has been proved to influence these results in a positive way.

In evolutionary change management, there's a tried and tested methodology for implementing this: the retrospective. It serves as an internal radar system that makes current impressions visible at a meta-level and distills them into conscious experiences. Many changes in all the relevant organizational fields can be determined across the 360° perspective of this radar. The retrospective is an attempt to move away from the operational business and reveal the bigger picture behind the individual WiP limits, service classes, and service-level agreements. This picture is of course not going to be discovered by one person alone, but rather through shared discussion and interaction.

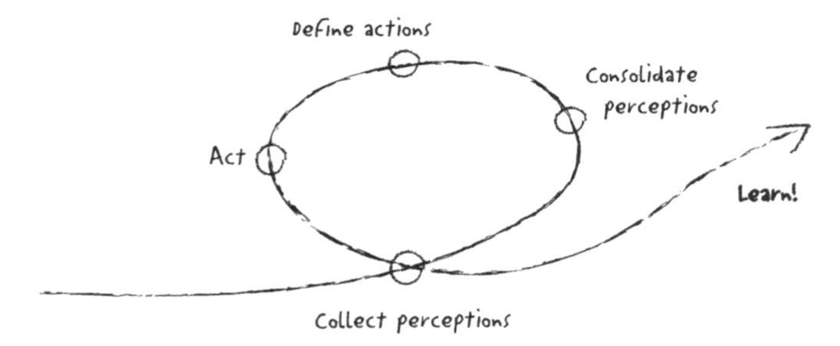

FIGURE 13.4 The systemic reflection loop.

This kind of reflective discovery demands one thing from everyone: the ability to pause or, more precisely, to pause in order to allow a review to take place and draw conclusions for further approaches. The so-called systemic reflection loop provides a simple and effective guide for this. As Figure 13.4 shows, this loop moves:

- *From* the **collection of individual perceptions**: "As far as I'm concerned…" and "What's been going on recently?"
- *Through* the **shared consolidation of these perceptions**: "What do we all perceive in a similar way?" and "What do we perceive differently?"
- *Through* the **recognition of characteristic patterns and the learning from these patterns**: "What picture emerges from this?" and "What conclusions can we draw?"
- *To* the **definition and implementation of follow-up actions**: "What do we want to try out in the future?"

Real progress brings with it a new collection of experiences used to create a further reflection loop.

Considered continuous movement, these reflection loops are primarily about observing ourselves as an organization for every single section of the value chain. To avoid reducing complex relationships to mechanical sequences, existing relationships should be inspected from a bird's-eye perspective, as it were. In his book "Managing," Henry Mintzberg strongly emphasizes the necessity of such inspection. It is precisely due to the dynamic of both market and organization that regular time-outs are necessary. Management must also repeatedly take a step back from the daily workflow and complement the frequent leaping into action with conscious reflection. In his own provocative words, "reflection without action might be passive, but action without reflection is thoughtless" [15]. Only this conscious stepping out of the flow of operational processes guarantees the kind of culture of mindfulness that is so important for every organization in this day and age.

In Kanban, individual perception is therefore inseparably linked with communication in its various forms. Alongside retrospectives, team exercises, leadership workshops, and method training courses are established formats for driving reflection forward. As already mentioned, Kanban draws on two reflection loops more than any others: a short one in the course of the stand-up meeting and a long one in the course of the operations review. In the first phase of the change process, the interview format helps to cultivate another very important loop. By way of semistructured, one-on-one, or group conversations, all relevant stakeholders are invited to take part in joint reflection on the status quo and the envisioned change at the beginning of the improvement journey.

As in the case of the Kanbanization of one telecommunications organization, such interviews can sometimes work real wonders. In this particular case, during an initial conversation with the CTO and HR, the situation was presented to us as messy. Apparently, the relationship between software development, operations, and product management had been fairly tense for quite some time. IT had been working on a huge load of problems for more than a year, and it didn't look like they could be sorted out. The large number of bugs triggered a series of complaints and caused dissatisfaction in the business areas, which by then had become chronic. Within IT itself, this reinforced a culture of finger pointing, which certainly hasn't helped solve the problems. Interviews were conducted with key players of IT as well as some of the relevant stakeholders of the business, which provided not only an in-depth reflection of the problematic situation but also the possibility to express emotions, vent anger, and voice concerns, creating the basis for solution-oriented thinking, which subsequently became a core value for the group workshops. Step by step, kaizen was attained all the way from individual conversations, through the shared consolidation of the most important themes, right up to a sequence of shared, immediate measures that were agreed upon.

Whether this reflection brings the desired results is of course once again a question of corporate culture. Kaizen is not an application that can simply be loaded in a system and then used as and when required.

If trust is absent from teamwork, if the general attitude is not oriented toward improvement, and if the daily forms of behavior are not team oriented, then even the best form of communication in the world will be of no use at all. If such a culture is in place, joint reflection would be of great help in the processing of the more delicate aspects of change. Fears and disappointments can be expressed; exceptions and resistances explored; worries and concerns brought into professional discussions. This is how an open culture of reflection helps Kanban change agents discuss as early as possible things that could block or at least slow down change.

13.2.2 The Power of Dialog

Together with critical self-reflection, it is above all cultivated dialog that has the power to free up undreamt energy. But what is dialog? And what is it that makes it so powerful? Communications expert William Isaacs considers dialog a dynamic connection between people, similar to a river flowing between its banks. As "the art of thinking

LISTENING	RESPECTING
Saying what I really think, what I feel and what I question.	Being calm on the inside, being attentive, feeling out differences.
SUSPENDING	**VOICING**
Holding back my own assumptions and assessments and questioning the obvious	Acknowledging differences and turning them into the source of new experiences

FIGURE 13.5 The art of dialog.

together" [16], dialog is a discipline that promotes personal, team-based, and organization-wide learning to equal extents. As Figure 13.5 shows, this discipline is based on four abilities:

1. **Listening**: the ability to consider other perspectives without overloading them with your own views. Listening is important here, as is the observation of nonverbal signals and moods.
2. **Respecting**: the ability to acknowledge conversation partners for what they are.
3. **Suspending**: the ability to take a step back from evaluations that are often simply too hasty and allow space for other perspectives.
4. **Voicing**: the ability to say what you truly think. Our own experiences suggest that this includes the capability to ask the right questions.

The dialog developed by Isaacs is built on many well-known elements. Nevertheless, its implementation requires effort. Dialog is provocative, applying forceful deceleration to today's manic high-speed life, conscious restraint to hurrying ahead, and thinking together to hero-like individualism. Dialog needs people who are capable of being surprised by what they say, who haven't already arranged all their thoughts, and who are prepared to let themselves be influenced by the conversation. That fits just as well with the necessary cultivation of attentiveness as with competence in shaping change gradually together instead of simply announcing or even decreeing it.

In his book *Getting Change Right*, Seth Kahan also identifies dialog as a fundamental element of any change initiative [17]. According to Kahan, successful change is promoted by the cultivation of relationships. "Interaction lays the groundwork of the future. You must get out of your office and talk to people" is his advice for any change agent. It is only with this form of dialog that the basis for a powerful setting in motion is created, and this is one of the crucial drivers of all change.

13.3 PROCESS DESIGN

If we identify dialog and reflection as fundamental elements, then we need to ask ourselves how we can incorporate them in our tailor-made Kanban change management. There is also the question of the guidelines we should build on for such an initiative. In our training and consultancy practice, we utilize a simple four-phase model for this. Each phase targets specific goals that can be achieved through selected design elements and change tools. The following are guidelines for the introduction of continuous improvement:

- **The phase of general clarification**, in which you gather all your first improvement impulses and mark out the personal, content-based, and organizational boundaries for your Kanban initiative
- **The phase of deeper understanding**, in which you identify your most important stakeholders, determine their perspectives in personal conversations, and create a consolidated catalog of demands for improvement
- **The phase of system design**, in which a group of selected key players build up a Kanban system tailor-made to the catalog of demands and establish agreements with all relevant stakeholders in order to transfer Kanban into the daily business
- **The phase of operation**, in which you do everything possible to foster a culture of continuous improvement

These four phases correspond to the architectural process of building a house: they establish the rough framework for the process of change within which design elements of all sorts can be used. To continue the building analogy, it is within the architectural parameters (walls, staircase, rooms, doors, and windows) that the concrete furnishings are installed (carpeting or wooden flooring, fitted or loose furniture, ambient or spot lighting). Architecture and design in change management always consist of five dimensions of intervention [18]:

1. **Factual dimension:** Which themes must be tackled? What do we need to focus on? Where are the critical pain points?
2. **Social dimension:** Who needs to be involved? When and in what way? Who are the stakeholders? Who are the key players in the planned change? Who can I leave out without having repercussions? Who do I need to inform about what?
3. **Time dimension:** How much time has been assigned to each individual change step? What progress needs to be made within any given time frame? What needs to take place more or less simultaneously?
4. **Spatial dimension:** Where is the process of improvement taking place? What space has been created for it? How much room for maneuver is there? What possibilities are there for retreat and informal action?
5. **Symbolic dimension:** Which models are used? What is the change? How is it staged?

The concept of architecture prompts thoughts of static conditions, but in actual fact, Kanban's framework is as flexible as each change step within a certain phase. There are of course a few fundamental principles: you shouldn't build your house on sand but rather on firm foundations; you should consider the division of the rooms; and you should not forget windows, doors, or roof so that you're not standing in a blacked-out prison or under a rainy sky at the end of the day. However, fundamental architectural rules enable flexibility: you can relocate walls, reshape rooms, expand the attic, add extensions, buy the neighbor's plot of land and expand, dig a cellar, add a floor, and so on and so forth.

There are also many options for the level of design elements: you can furnish the same room with carpeting or laminate flooring and pack it full of dark furniture or go for a lighter touch; you can paint or wallpaper and, for a little more money of course, change the red-and-blue-striped wallpaper for a blue-and-green polka-dot one, at any time.

The same goes for organizational Kanban change leadership you can base the general clarification on a checklist-style run-through or just as easily do it through dialog; you can retain the role of initiator or sponsor or take over the role of coach or facilitator; as we will see in the next few chapters, you can draw on the most varied means of support in order to prepare yourself for your role; and you can chuck all the information you gathered in the course of the problem analysis into a spreadsheet, work with a slideshow, or make do with Post-it notes that you place in front of the participating stakeholders, "putting your cards on the table" as they say. Spontaneous breakout sessions are just as possible as plenary discussions. Results can be presented centrally or exhibited in the form of a gallery. You can grant specific topics more time in the last minute or alternatively insert a break to get some fresh air.

Due to the flexibility of the organizational design, it will hardly come as a surprise that you also have many possibilities for the actual tools of change. Using selected case studies, we will present these possibilities in detail in Part 3 of this book and demonstrate how the introductory process varies according to situation and corporate context.

In conclusion, we can definitively say that every Kanban change management initiative needs builders, designers, and skilled people.

These roles can be held either by one person, distributed among several people, or delegated to external experts.

"Do your own thing!" says the TV advert for a DIY shop in which a German comedian's DIY heart sinks at the sight of a totally run-down house. "Every change needs a beginning," says another DIY concern, summing up its DIY philosophy, and concludes, "There's always something to be done." Consequently, people hammer around with grumpy expressions as if there was no tomorrow, while others drill with clenched jaws or break through walls with brute force. Managers have the habit of starting out in their change processes just as determinedly and doggedly: they plan and design, announce and promise, roll out new things, and convert others. For each of these steps, there is always a good change tool at hand.

The saying "a fool with a tool is still a fool" is as true as ever today in change management as anywhere else. Consequently, we are particularly keen to close this chapter with a few remarks on the subject of attitude.

Before we come to the real case studies of evolutionary change management using Kanban, we would urge you to:

- **Respect the autonomy of social systems.** The tenacity of organizations means that change can be shaped to a certain extent, but it can never truly be planned.
- **Have faith in the capacity of organizations to help themselves** as well as people's willingness to mobilize this force.
- **Acknowledge people affected by change.** Their actions are also reasonable when they create resistance and won't get any more reasonable if we try to break this resistance.
- **Use the principle of many pairs of glasses** because the world is too colorful for a single perspective and can only be comprehended from various points of view.
- **Understand intervention as a catalyst** whose impact can never be foretold.
- **Realistically evaluate the true complexity** of the intended change and estimate the time it will take to process this change productively.
- **Move away from the heroic ideal of self-made man and solution hero** who has both organization and change under his control.
- **Nurture the ability to use your own ignorance, confusion, and uncertainty as resources.**
- **Strike a good balance between self-awareness and humility** in order to appreciate your own contribution to successful change without overestimating it.
- **Have fun in learning and in the kind of continuous improvement** that is fed precisely by your own mistakes and shortcomings.

Having made this wholehearted plea, we will now embark on our journey into the world of practical Kanban change leadership.

WHAT YOU CAN TAKE AWAY FROM THIS CHAPTER

In the twenty-first century, change management builds on three core competences: mindfulness, professional communication, and agile design.

The market and the organization itself must be taken into account. Structures and processes, the culture of collaboration and self-reflection, and above all organizational leadership play key roles.

An appropriate understanding of management capabilities for the twenty-first century favors personal role modeling and the consistency of leadership activity. This tightly couples each attempt to improve organizational processes with management improvement and brings "leadership as a team sport" to the fore.

Credible answers to the three core change questions "*must* we?" "*can* we?" and "do we *want* to?" must be communicated. Professional dialog is a core element of this kind of communication and is based on respect, honest statements, personal restraint, and active listening.

The change process itself progresses through four typical phases:

1. General clarification
2. Deeper understanding
3. System design
4. Operation

Each phase targets specific goals that can be achieved through specific change management tools. Along with craftsmanship and know-how, the right mindset plays just as important a role in the cultivation of continuous improvement.

PART 3

KANBAN CHANGE LEADERSHIP

14

FROM THE IDEA TO THE INITIATIVE

Great sportsmen invest a huge amount of time in their training regimes and much less in the decisive competition itself. The focus of all exceptional athletes is on preparation. This is the phase when the essentials for the whole season and the foundations for the desired success are laid. According to American management experts Jim Loehr and Tony Schwartz, it is the other way round with managers: they barely train at all and spend their whole time in competition instead [1].

The fact that managers train very little takes its toll—primarily in the nature of the competition. Twenty-first-century markets are turbulent and constant change is inescapable. This turns leadership into an energy-draining sport and makes intensive preparation essential for successful change management.

However, necessary changes are nevertheless frequently neglected, along with the essential training for their implementation. Decades ago, Peter Drucker commented that everybody "has accepted by now that change is unavoidable. But that still implies that change is like death and taxes—it should be postponed as long as possible and no change would be vastly preferable" [2]. His observation is as valid for many organizations now as it was then. The fact that changes are put off is related to the uncertainty, anger, and fear associated with the changes themselves. As experience shows, the procrastination bug is also associated with a profound lack of knowledge and practice. Without good organizational training, changes are more suffered than implemented.

Kanban Change Leadership: Creating a Culture of Continuous Improvement,
First Edition. Klaus Leopold and Siegfried Kaltenecker.

What do we need to watch out for in particular? What are the most important laws for Kanban change management? How can we create a culture in which continuous change steps are a part of daily business? These questions highlight the training grounds we are going to describe in detail in this part of the book. Using real case studies, we lead you step-by-step from professional preparation to effective implementation right through to sustainable cultivation of continuous improvement. We don't want to highlight only the factors for success but also the stumbling blocks—such as dealing with resistance, coping with change conflicts, or engaging in goal-oriented communication with all stakeholders—that repeatedly appear in real-life practice. Of course, overcoming these traps and difficulties always depends on the specific situation: the people involved, the team dynamic, and the corporate context.

We therefore cannot deliver a magic formula that you need only copy–paste to your organization. But we are more than prepared to show you all sorts of practical approaches and methods that have proved their value in various Kanban change situations.

These tried and tested practices can be grouped together into four typical phases:

1. **General clarification**, in which you define the starting point, assess Kanban as a possible solution, and specify the contextual conditions for further progress
2. **Deeper understanding**, in which you gather the perspectives and interests of your most important stakeholders and condense them into a specific catalog of root cause-analyzed problems and demands for further improvement
3. **System design**, in which you translate this catalog into a tailor-made Kanban system and get this system up and running together with your stakeholders
4. **Operation**, in which you do all you can to propagate continuous improvement

We've dedicated an entire chapter to each phase, drawing on our wealth of experience as consultants and trainers to provide relevant, real-life case studies. The following is an overview of the people and situations you will come across time and again in the coming chapters.

CASE STUDIES FOR THE APPLICATION OF KANBAN

- **Roswitha Münz**, COO Request and Change in an insurance company; 41 employees in her department; about 900 in the entire organization
  - **Why Kanban?** To deliver high-quality processes in the entire area of responsibility and ensure permanent improvement
- **Josef Drechsler**, Head of IT Department in an infrastructure organization; 8 employees in his department; 800 in the entire organization
  - **Why Kanban?** Huge external pressure; permanent intervention from all sides; lack of clarity in processes

- **Helga Rösner**, Head of Development for a media organization with 600 employees; 22 in her department
  - **Why Kanban?** Limited success of previous change management projects; conviction that an evolutionary approach has a greater chance of success; need of improvement in terms of coordinated progress
- **Susanne Schweizer**, Head of Development in a medium-sized pharmaceuticals firm; 9 employees in her department; 250 overall
  - **Why Kanban?** The team often does low-priority work that also takes up a lot of time; product management has other ideas; management has a tendency to cut right across things and micromanage; the team lacks clear parameters.
- **Thomas Müller**, Head of Payment and Projects Department in an international energy firm with more than 30,000 employees; 18 employees in his department
  - **Why Kanban?** It enables the reshaping of out-of-date work processes while simultaneously modeling business affairs.
- **Herbert Krakauer**, Head of IT Department and member of the general management of an organization that specializes in security solutions for hardware and software with 250 employees in total; 14 in the IT department
  - **Why Kanban?** Periodic overload; constant reprioritization; repeated disruptions and breaks in the workflow; latent customer dissatisfaction
- **Stefan Bergmüller**, Team Leader for Second-Level Support in a financial services organization; 7 employees in his team; 2000 overall
  - **Why Kanban?** Chronic overload; low esteem; lack of understanding; a thousand things to do at the same time; lack of prioritization

As already mentioned, all these examples are taken from the real working world, the world in which we were involved as Kanban coaches, change consultants, or facilitators for a period of time. Organizations have been kept anonymous, and names have been changed for data protection reasons.

At the end of each chapter, we summarize the most important points you need to consider for your personal training program. Regardless of the fact that this change management is to be understood today as a team sport rather than the performance of an individual, every improvement work begins at the individual level. In other words, it begins with you.

The program that trains and prepares you for successful change management is replete with substantial questions: Why change at all? Why now? Why with Kanban?

These are typical questions regarding the need for change. Every change agent must answer them. The same goes for the definition of the capabilities required for the change: What precisely to we need? What knowledge do we need to construct or

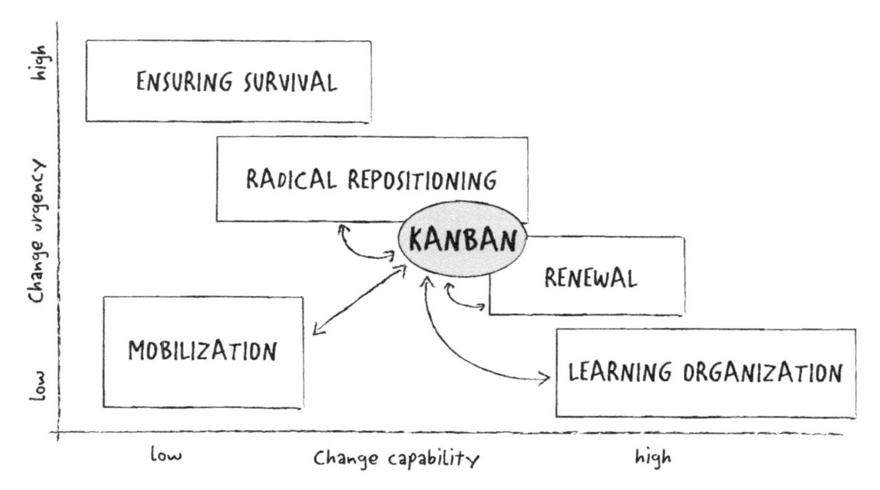

FIGURE 14.1 Kanban on the change map.

develop: of Kanban, of change, or of management? Which capabilities do we need to hone to be able to achieve our goals? Which of these can we acquire ourselves and which will we need external help with?

As made clear by Barbara Heitger and Alexander Doujak's change map [3], as shown in Figure 14.1, the application of Kanban is most productive in cases of a low to moderate need for change. If there's a lot of pressure, customers openly show their dissatisfaction, teams are already grumbling loudly, or the entire department is up with its back against the wall, more drastic measures are perhaps required. If there's less pressure and processes run smoothly, then it is tempting to be content with more basic measures for mobilization.

If you have chosen Kanban, training is required. Just because an evolutionary approach to change is being implemented doesn't mean that the bar is set any lower than with other approaches. The key objective, namely, that of creating continuous improvement, provides a degree of complexity that is by no means inferior to "revolutionary" approaches. Whoever is involved with culture—and in terms of Kanban, this necessarily means you—will open a Pandora's box, so to speak.

We advise care with this box. An approach where you simply get stuck in with Kanban can easily backfire. Ultimately, it's not about quick fixes but rather the creation of a sustainable culture of improvement. "It shows who is committed for good," said a colleague of ours on the subject of how long Kaizen takes. So be sure to take stock before you make such a commitment. Define your starting point using our guidelines before you begin on your Kanban adventure. And create a training program that is tailor-made to your personal working situation from the exercises we offer.

Good luck!

15

GENERAL CLARIFICATION

Have you been to a lecture about Kaizen and had your interest roused? Through reading about Kanban, have you come to the conclusion that you want to continuously work on improvements? Have you just returned from a lean workshop, all fired up for cultural change? Good. Then you've got the fire burning in you that's crucial for any successful change initiative. But, so you don't get consumed by your own flames, we urgently recommend that you pause for a moment. Before you set to work on the actual implementation, you should take a minute to calmly contemplate the starting point: check the context of your change impulses; go over your own options; get a better idea of what it is you're really getting yourself into; and get yourself into good shape for the organizational clarification needed to translate your change ideas into a good initiative. The following list of questions helps you in this personal reflection.

LIST OF QUESTIONS FOR KANBAN INITIATORS

- Why change?
  - Are we being forced into it by external powers?
  - By changes to our context? By a crisis?
  - Or primarily by the desire to create something new?

Kanban Change Leadership: Creating a Culture of Continuous Improvement, First Edition. Klaus Leopold and Siegfried Kaltenecker.

- Who is interested in this kind of change?
  - For whom is the change a major concern?
  - Who are the people already championing continuous improvement?
  - Who might have objections? Who might even resist?
- What are our goals?
  - How will we know that the change has been successful?
  - What's in it for us, when we change ourselves?
  - How will our customers profit from our efforts?
- Why Kanban?
  - What are the distinguishing features of this type of change management?
  - How does Kanban help us achieve the desired improvements?
  - What do we need to do in order to optimally apply evolutionary change?
- What do I personally need to do?
  - How do I best implement my ideas in the organizational context?
  - What should I particularly prepare for?
  - What will I do next?

Let us present our first case study to show you the answers you can get working through this list of questions. Herbert Krakauer is a team leader in a very successful Swiss organization that specializes in security systems. Apart from being the operational leader of a team of 14 software and hardware professionals, Mr. Krakauer is a member of the strategic management. He hears about Kanban for the first time from a CTO friend of his. "Achieve significant improvements with manageable effort," is the core message from this friend, who has by now implemented Kanban in his entire IT department. Intrigued, Mr. Krakauer follows his friend's advice to set aside a little time to clarify where he should start from. These are the key findings of his self-inspection:

- In Mr. Krakauer's opinion, change is needed mostly due to the chronic overloading of his team. In daily business, there's simply too much being loaded onto his people: constant reprioritizations via urgent implementations, permanent disruption, frequent delays, and greater pressure from some of the customers all increase the feeling of not seeing the wood for the trees. Despite economic success, this "permanent state of alert" (in the words of one team member) leads to a latent dissatisfaction in the team and of course isn't always great for the customers.
- Mr. Krakauer is convinced that many people are interested in improving things. True, some reckon that the situation "is just what it is," but he believes the majority would definitely support a reasonable initiative. This goes for his team and also for sales and marketing, second-level support with which a structured knowledge transfer had indeed already been considered, as well as other

members of management. Moreover, it's hopefully also the case for the customers, who could be won over by significant improvement to the lead times as a result of Kanban.

- Mr. Krakauer believes that Kanban will be seen as a particularly positive approach because the change will be implemented in small steps. The team has neither the time nor the will for a large change project. Mr. Krakauer still hears his friend's words in his ear saying that an increase in efficiency and higher flexibility are also achievable with an evolutionary approach to improvement. More than anything else, visualizing the actual work processes and securing a better way of processing orders via the introduction of WiP limits are highly attractive outlooks.
- In terms of impact, Kanban promises Mr. Krakauer one thing above all: significant relief. Firstly, jobs should be done in the right order, and secondly, they should be more accommodating to the needs of the customer. Mr. Krakauer believes he stands a good chance of creating a significantly higher degree of reliability through the introduction of classes of service.
- Mr. Krakauer also wants to use Kanban to create a structure for better communication. The creation of a shared board and the introduction of regular meetings offer the possibility to keep an eye on all processes. In terms of new capabilities, good facilitation skills are needed more than anything else because otherwise the team is liable to get bogged down in details.
- In order to be a strong Kanban change manager, Herbert Krakauer personally undertakes three things: firstly, a deeper study of Kanban; secondly, an initial meeting with the Kanban coach who supported his CTO friend in the kickoff phase of his change; and, thirdly, a more detailed discussion within the management team in order to seek a definitive mandate for the introduction.

15.1 CLARIFY THE METHOD

Without a proper understanding of the method, Kanban is like the proverbial pig in a poke. Who wants to buy it without seeing it first? Therefore, we recommend that you clarify as soon as possible and as broadly as possible what Kanban is actually about, what evolutionary change management is, and how this approach can help you to significantly improve your specific organizational situation. Because without this kind of foundation, you will have a hard time convincing others of the need for continuous improvement, let alone succeed in it.

A compact Kanban introduction shores up the knowledge foundation on which you can further build your culture of continuous improvement. Let's look at Roswitha Münz's case in order to identify the value of this kind of basic training. The COO of an insurance company for many years now, Ms. Münz knows that the design of her improvement initiative is dependent upon her corporate culture. Her approach is determined by the history, size, branches, and environment of her organization as well as the experiences of change it has had in the past.

Nevertheless, there are a couple of general guidelines for successful Kanbanization. Ms. Münz has taken away three strong markers from her reading of David J. Anderson's book [1]:

- **Kanban starts with what you do now**. No structural reorganizations are necessary, no expansive training programs or process revolutions. After several experiences with drastic *Business Process Reengineering* projects, Kanban's approach instantly seems attractive. She can simply identify her current work processes and take some initial improvement steps.
- **Kanban respects the existing order**. Neither the existing processes nor responsibilities or job titles are necessarily called into question. Fittingly for the conservative environment in which Ms. Münz operates, an evolutionary approach seems to her more promising than a revolutionary one. Instead of alienating her business partners with radical measures, she will start with what is currently in place. Kanban makes it possible to convince the partners to participate in an improvement project that successively increases the existing value.
- **Kanban pursues incremental, evolutionary change**. Due to her previous experience with organizational change, Ms. Münz has more faith in a step-by-step approach rather than one great leap. Agreement with everyone essentially involved in the change movement is required for this.
- **Kanban needs leadership at all levels of the organization**. It is clear to Ms. Münz that all those involved must contribute their own ideas for improvement if the continuous improvement is to work. The operationally active employees know best what needs to be improved in their day-to-day work. Accordingly, Ms. Münz intends to use this expertise better than in the past in order to take the right improvement steps *together*.

Even after the initial preparation, it is clear to Ms. Münz that she needs to get as many of her stakeholders as possible to attend an introductory workshop on Kanban. "The more we manage to create a shared knowledge base on Kanban, the better," she says, explaining what she intends to do. We agree with her that Kanban requires a common understanding of work and improvement. The introductory workshop "Kanban for decision-makers" that we offer is the perfect solution. This workshop provides an overview of everything that is important for a practical implementation of the principles established by David J. Anderson: visualization, WiP limits, flow, risk management, classes of service, and metrics. The core knowledge about these is combined with practical exercises in which the acquired knowledge is applied to the specific situation. Ms. Münz describes this as "cookery instructions that provide us with information about the kitchen, kitchen utensils, and possible recipes as well as providing us with the actual ingredients."

Taking up this analogy, we should explain that in the basic training we are above all concerned with understanding the "what for."

Why do we actually cook? What do we achieve by doing it? Who benefits? Translating this into real terms, alongside Kanban's toolkit and various application exercises, the frequently mentioned big picture is also important in basic training. Real-world, practical implementation of evolutionary change management shows us that a common understanding of the "what for," the "what," and the "how" of change is of great importance. We believe that the following guidelines are significant:

- **Kanban is a change initiative**. It is concerned with systemic improvements for which it is not the performance of individuals but rather teamwork that is crucial. Value creation and quality of work both increase due to the clear rules for all relevant stakeholders.
- **Kanban is about the work culture as a whole**. Improving this culture requires regular reflection on one's own fundamental actions—inspection that expresses itself with a particular attitude to performance and collaboration. This then requires the willingness to work diligently at your own development.
- **Kanban is centered on the people and not on the mechanics**. It is people that drive sustainable improvement, and they do this very identifiably through emotions: joy, courage, enthusiasm, but also anger, disappointment, and sadness. We recommend that you respect these emotions and use them well since they can ultimately be considered the catalyst of change.
- **Kanban is a team sport**. Ms. Münz feels that this point is particularly pertinent. She needs strong allies in order to create a culture of continuous improvement. Together, we define this basic training program as an important step toward bringing change to life and, together with Ms. Münz's business partners, laying the foundations for sustaining this change in the long term.

Establishing the broad framework is very much part of Ms. Münz's thinking. Always one for a good metaphor, Ms. Münz compares this framework with a panorama lens that helps to position tools and process components in the right relationships. It becomes clear that we cannot explore everything in a 1-day training course, but the bigger picture helps to put our Kanban foundation on the right track.

15.2 CLARIFY THE ORGANIZATIONAL CONTEXT

"If you believe that nothing works anymore, then an impulse for change will appear from somewhere," goes one gem of pop philosophy, somewhat loosely interpreted. Every organization's development is driven by such impulses. But the thing is, you only get change off the ground once the organizational framework has been ascertained, along with the personal mindset and your understanding of the content.

Kanban also needs this kind of clarification. The idea of evolutionary change is driven by the principle of small steps. These steps take a hold at the organization's current position and initially leave the work processes, job titles, and responsibilities completely untouched. "Everything from which the team members and other partners, participants, and stakeholders derive their self-esteem, professional pride, and ego should be left unchanged," says David J. Anderson [1, p. 63.].

But even if as little as possible is changed at the beginning, Kanban represents a systemic intervention that alters the existing network.

Such an intervention shouldn't take place without reason. Neither should you intervene without having voted on it as a group. "Contract" is the technical term for this voting procedure, which deals with concrete expectations and explicit agreements about further progress. Together with your employees, colleagues, and superiors, you come to an understanding with respect to the reasons, goals, and conditions for your Kanban initiative. And here, the guidelines go hand in hand with the definition of the "guiding coalition" that change guru John Kotter [2] establishes as the social basis of every successful change initiative. Who is managing the initiative? Who takes on responsibility and for what? What particular resources are necessary? And what do the individual implementation steps look like?

THE KANBAN CONTRACT

Our experience shows that, more than anything, the following points should be clarified in order to establish a solid organizational foundation for your Kanban initiative:

- Why should our organizational unit change?
- Why should we change now?
- Why should we use Kanban to model this change?
- What do we expect of the Kanban introduction?
- What is our estimation of the costs?
- Who is responsible for the change process?
- What roles, competencies, and resources are necessary?

We strongly recommend that you work through these questions as comprehensively as possible before bringing them to your superiors. It is also worth preparing concrete references, dates from comparable initiatives, and some success stories for this contract.

"Guiding coalitions" can appear to be very different from one another. These differences are a result of:

- The organization's current economic situation
- The current situation of the value chain in which Kanban is to be applied
- Previous experiences with change initiatives
- The culture, hence also the power politics, that has a large effect on change management

- The hierarchic position of the person initiating Kanban
- The relationship, particularly the amount of trust and openness, between the contracting partners

Let us use a further case study to hone our understanding of the contracting process. For 2 years now, Helga Rösner has been Head of Development in a media organization with 600 employees. According to the debriefings, the organization is doing well; the work environment and the current atmosphere are also good. However, Ms. Rösner sees enormous potential for improvement in her own area of responsibility: the uncoordinated approach of the developers, their middling collaboration with the testing department, and the countless interjections from production in particular are all things that really get on everyone's nerves.

She first hears of Kanban at a conference in London—an expert's presentation on lean and agile management arouses her curiosity, she learns more about it, and her gut feeling is confirmed. Yes, Kanban is definitely the right option for her team!

She makes an appointment with her CTO in order to convert the idea into reality as quickly as possible. In order to give her proposal more gravitas, Helga Rösner has asked around in her management network and researched two *success stories*. She provides her CTO with three reasons why Kanban is necessary: firstly, the need to respond better to the highly dynamic environment in their organization; secondly, the possibility of significantly improving the application of available expertise through better processes; and thirdly, the evolutionary approach, which spares all participants complicated project negotiations.

Nevertheless, Kanban also needs clear rules. After having clarified the "we must," "we can," "we want to," and "we are allowed to" aspects of change, Ms. Rösner agrees the following with her CTO:

- A **fundamental commitment** to implement a series of small improvement steps rather than the previous, large project goals with fixed milestones
- **Personal reporting** on the initiative within their regular bilateral management meeting
- **Good alignment** between all team members and the most important partners in the development department
- The **deployment of specific metrics** that constantly provide information concerning the success of the Kanban initiative
- **Effective teamwork** between the CTO as sponsor and Ms. Rösner as Kanban change manager
- **A dedicated budget**, giving Ms. Rösner the ability to purchase specific Kanban and change management expertise

With these valuable arrows in her quiver, Helga Rösner gets ready to define the next steps. She quickly realizes that none of these steps are possible without deeper clarification of her own role in the process.

WHAT YOU CAN TAKE AWAY FROM THIS CHAPTER

As we all know, "practice makes perfect." Practice is also recommended if you want to apply Kanban as well as possible in your organization.

We recommend the following as the first things to do:

- Personal clarification of your starting point
- Content-based clarification of which powers you are planning on summoning using Kanban
- Organizational clarification of the boundaries within which you will introduce Kanban in order to establish a culture of continuous improvement

To this end, we've shown you the following change tools and their corresponding applications in this chapter.

Tool	Application
List of questions for Kanban initiators	Reflection upon your own starting point Double-checking the motives and goals
Kanban basic training	Content-based foundation for improvement An overview of the principles and practices Relevance for your particular starting point Ensuring a common understanding in order to make an informed decision
Kanban contract	Clarification of the organizational context and the specific expectation Explicit agreement of the boundaries of the improvement initiative

16

DEEPER UNDERSTANDING

In the style of a master steeplechase runner, you've cleared the first few hurdles and converted your first improvement ideas into a real Kanban initiative. However, now you're back at the beginning and have to settle the specific demands on this initiative in detail—together with those making the demands. The magic term here is "stakeholder management."

In this section, we will let you in on a secret—we will introduce several tools for stakeholder management that have proved themselves in diverse change initiatives:

- **The personal retrospective**, where you apply the principle "Start with what you do now" to yourself and commence the Kanban initiative with a more detailed specification of your own perspective
- **The team constellation**, where you answer the question of existing power relationships in your immediate area of responsibility, thus helping you clarify your own assumptions in terms of willingness to change
- **The change dialog**, with which you can prevent the onset of skepticism, worry, and fear arising from tangible conflict
- **The team conversation**, in which you can work constructively on a challenging starting point at the team level
- **The team retrospective**, which invites everyone to review what worked well, what did not work so well, and what did not work at all
- **The stakeholder map**, which visually defines your most important business partners and their relationships, created together in the group

Kanban Change Leadership: Creating a Culture of Continuous Improvement,
First Edition. Klaus Leopold and Siegfried Kaltenecker.

- **The stakeholder interviews**, through which you come into personal contact with your partners, explore their particular perspectives, and are able to identify the demands on your Kanban system
- **The stakeholder workshop**, in which you invite all involved business partners to a general evaluation of the interview results and finalize the demands catalog

16.1 THE PERSONAL RETROSPECTIVE

The personal retrospective provides a special opportunity to explore your field of change. We will follow the American pioneer project manager Norm Kerth in defining the retrospective as a catalyst for sustainable learning [1]. But this doesn't mean we're just looking at what's already happened. Looking to the past is valuable only if we are mindful of the future.

It is well known that if you don't know your past, you're doomed to repeat it. So how can you learn about your past? And particularly, how do you learn from your past mistakes, deficits, and blind spots? The credo of the retrospective is that learning takes place through reflection upon working experiences. One reflects on the flow of events in order to distill conscious experiences, such that strengths are identified and the repetition of known mistakes can be avoided.

As we will illustrate, the team retrospective is a strong driver of continuous improvement. However, in the do-it-yourself variation, we are initially concerned with your personal balance. Let us use the case study of Mr. Bergmüller to illustrate the value of this balance.

Stefan Bergmüller is the team leader of a second-level support unit in a major financial services provider. Through his extensive technical knowledge and practical experience gleaned from more than 10 years with the company, Mr. Bergmüller has become, in the words of his boss, "master of the entire server infrastructure." Unfortunately, this role comes with a huge workload. Sometimes, Mr. Bergmüller feels like everything, absolutely everything, goes through him. His team's daily operations have become as familiar to him as the feeling that he's the Don Quixote of support, tilting at the windmills that are the organization's IT problems. There's simply not enough time and things pile up relentlessly.

Mr. Bergmüller providentially gets hold of the guide printed below and decides to analyze his work situation as soon as possible.

PERSONAL RETROSPECTIVE

Context

- Guaranteed time window (in our experience, at least 20 min)
- Quiet location (usually away from your normal workplace)
- Pleasant atmosphere (somewhere you like being)

Core questions

- What has gone well recently? What's been successful in my area of operations? Which improvements have succeeded?
- What has not gone well recently? What failures have there been? What's gotten even worse?
- What conclusions can I draw from this balance? What should we absolutely hold onto in future? What do we urgently need to change? What do we need to talk about in more detail?

Approach

- Get hold of some Post-its and a pen.
- Start by writing all you can think of in answer to the first two core questions. We recommend that you begin by concentrating on the first group of questions and only going on to the failures once you've finished with your success retrospective.
- By writing down all the factors for the successes and failures on a separate card, you will be able to expand on cases or cluster similar cases together at a later stage.
- Once you've completed your retrospective brainstorming for the two groups of questions, look through your collection. Can you identify a pattern? Can anything more be said?
- Start your conclusions. Go through the third group of questions and identify the most important findings and incentives for improvement.
- To complete your retrospective, translate your conclusions into real action points and put these in a sensible order.

Typically, it's 3 days before Stefan Bergmüller has time to do his retrospective. Nevertheless, one evening, he gets around to doing it: he finds a table in a quiet corner of his favorite café, takes a pile of Post-its out of his bag, and begins with the brainstorming. Below is an overview of the answers that Mr. Bergmüller writes to the core questions:

- **Very good to good**: (i) Team spirit, (ii) mutual support, (iii) resilience, and (iv) praise from Franz (Head of Department, Cash Management)
- **Less good to not so good**: (i) Chronic work overload; (ii) lack of appreciation; (iii) lack of understanding on all sides; (iv) a thousand requests at once; (v) no external prioritization, "We have to guess what is really important because in our ticketing system everything is always super important!"; (vi) impatient customers; (vii) operational stress; (viii) no time for relief; (ix) the feeling of being

a hamster in a wheel; (x) the amount of sick leave; (xi) my people are getting burnout; (xii) I'm getting burnout; and (xiii) an increasing sense of lack of purpose (a propos Don Quixote).

- **Conclusions**: (i) It can't go on like this, (ii) enough with the hamster wheel, (iii) I need to get out or I'll have a breakdown, and (iv) I need to talk with Rudi (CIO, Mr. Bergmüller's boss) tomorrow.

It will come as no surprise to hear that the conversation with his boss didn't happen the next day but rather a whole week later. The case interests us because Kanban only emerged as a possible solution—indeed, one that was urgently needed—in the course of the conversation between Mr. Bergmüller and his boss. After one more week of full operational escalation, Mr. Bergmüller was a hair's width from throwing in the towel. It is probably only due to the good relationship between the CIO and the team leader that this didn't happen. With the support of an experienced Kanban coach and the blessing of the CIO, the decision was taken to "do everything possible so that by pooling all strengths, the course can be changed." As planned, the coach was contacted; as planned, the conversation took place; as planned, the team retrospective, defined as the first step, also happened. That all this was done took Mr. Bergmüller by surprise. Might the improvement actually stand a chance? Might the introduction of Kanban perhaps even be the precursor to continuous improvement?

Of course, the personal retrospective isn't only beneficial in sounding out the current situation. Experience tells us that individual preparation before the team retrospective is always beneficial, particularly if, as a manager, you are thinking of facilitating the team retrospective yourself. Through this preparation, you achieve an initial evaluation of what you think is particularly important. There's also less danger of getting sucked into the team's work processes.

A manager we know well recently reported an interesting variation to this preparatory work. Helga Rösner, Head of Development in a media organization with 600 employees, used this template for a retrospective in her last change initiative. She however changed the core questions slightly:

- What positive experiences have we had in the course of the change project? Which changes were successful?
- What didn't work? In which areas were we unable to implement our plans as we wanted? And where did the improvements fail completely?
- What are the things I'm still unsure about?
- What can I learn from my positive/negative/unsure balance? What would I definitely like to do differently in the next change initiative? What would I like to keep? And in which areas would I like to carry out further research?

With the assistance of these guiding questions, Helga Rösner was able to clarify her personal balance and start Kanban on the basis of actual learning experience. "Given our highly agile environment, we won't simply place Kanban in an untilled green field," emphasized Ms. Rösner. "I want to take previous experience with change into consideration. Otherwise, for some colleagues, it will simply be one more wave that they surf just like they have always done." With her personal retrospective, Ms. Rösner has shown herself that her change experiences "weren't just a walk in the park" in her area of responsibility. In order to fit everyone into the crowded boat of change, she resolves to hold a team retrospective or, to put it more precisely, a comprehensive review of previous change experiences with her key player.

16.2 THE TEAM CONSTELLATION

After the personal sounding out of your field of work, it is logical to proceed with a deeper examination of the protagonists in this field or, more precisely, with your ideas of the reactions Kanban will trigger in each individual. You can visualize these assumptions with the assistance of the so-called team constellation.

> *What is this about, then? Fundamentally, with the constellation method, we are trying to achieve a better picture of a certain situation. One's own assumptions about team members' willingness to change are also projected to a certain extent. The team members' suspected attitudes concerning Kanban are expressed in the constellation.*

You can do this by defining a specific object as an avatar for each and every team member. These proxies can be parts of a model kit: they could be Lego bricks, coins, buttons, or figurines—whatever you currently have to hand. The next step in your constellation is ordering these selected proxies into what you believe is the correct sequence—not just to each other but also with reference to the change goal.

Figure 16.1 shows the layout of one such team order. It shows the situation from the perspective of Susanne Schweizer, Head of Development in a midsized pharmaceutical organization, in the course of individual coaching during which she is preparing herself for the coming change measures. Ms. Schweizer has selected wooden bricks of various sizes and colors as her constellation avatars, each brick representing a particular employee. All the bricks are ordered as Ms. Schweizer sees fit, with reference to the change goal, that is, the "effective Kanbanization," and to each other. This requires:

- The **courage** to let yourself be open to such an experiment
- The ability to observe the familiar everyday from a **bird's-eye perspective**
- **Faith** in your own gut feeling
- The willingness to really tackle your **own observations** of your everyday working life
- The **flexibility** to try out different positions and relationships by literally grasping hold of them

At the end of the constellation, our Head of Development allocates each avatar a characteristic statement regarding change (the speech bubbles).

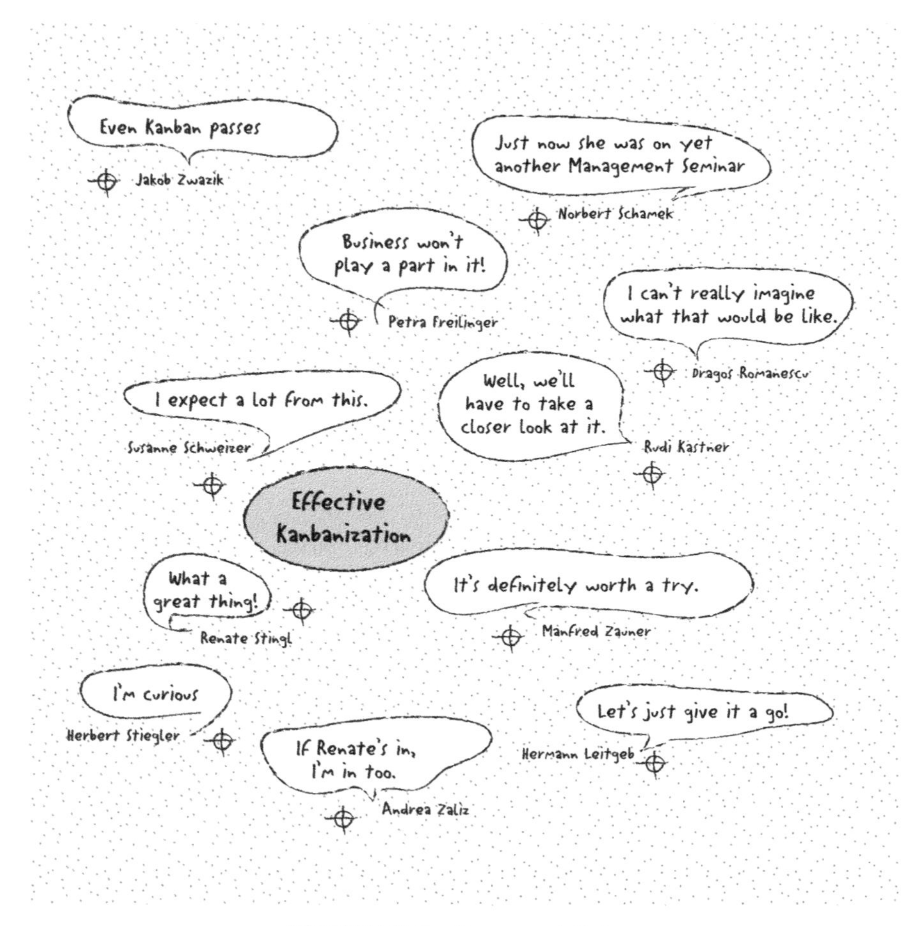

FIGURE 16.1 Team constellation.

What insights can Susanne Schweizer gain from this image?

1. Ms. Schweizer found the work process itself was interesting. "Which brick do I choose for which person? Whom do I consider first; whom do I see as last? How do I position everyone with reference to the goal but also with reference to all the others? Where does the image feel instantly right? Where does it not? What am I still unsure about, even at the end?"
2. Susanne Schweizer benefited from what she calls the "recognition of implicitness." Of course, according to Ms. Schweizer, she knew all this all along, or at least sensed it. But the constellation allowed her to give a voice to her tacit knowledge: "Ah, that's how I see that!" Some things came as a surprise to Ms. Schweizer. "I wasn't aware it was like that." Through her bird's-eye perspective, our Head of Development can gain new insights into what she only ever

partially experiences in the hectic rush of daily life. Susanne Schweizer summarizes it as follows: "It is the sum total of my experiences with this team and an estimate of how I believe every individual will react to my initiative."

3. Ms. Schweizer uses the constellation to make room for her gut feeling. What dynamic will the slogan "effective Kanbanization" trigger? And what should Ms. Schweizer be aware of if she wants to manage the change process as well as possible?

Together with the assumptions on how the various individuals will behave, the constellation highlights three subgroups and one outsider. The first group—to which Susanne Schweizer also sees herself as belonging—welcomes the coming change, even if it's just for methodological ("huge potential for improvement") or social reasons ("I'm in favor of it because my friend's in favor of it"). The second group is very much open minded but still skeptical ("we need to take a closer look at this"). The third group seems reserved or even dismissive ("it won't do any good anyway"). And then finally, there's the outsider, who is perhaps only waiting for the change wave to roll over him and away again.

All in all, the constellation says a lot about the mindsets Ms. Schweizer believes to be in place. However, beyond the purely individual behavior, the image also highlights the systemic forces for and against change. Establishing herself as initiator, Ms. Schweizer can on the one hand identify her allies: she expects a lot of positive energy for the Kanbanization from the group of young employees who are also personally close to her. On the other hand, she can also see the areas where she expects resistance: it's more likely to be the older employees, those who have been with the company for a long time, the "old school technicians." Last but not least, we have our outsider, who seems to be a lost cause right from the word go.

Of course, such a constellation doesn't generate an objective image of reality but rather a highly subjective representation of Susanne Schweizer herself. But it is precisely this subjectivity that allows our Head of Development to take a more detailed look at her own hopes and fears and therewith create a foundation for mindful change management that is ultimately always based on assumptions concerning:

- **The environment:** Why do we need to change?
- **The system:** How can we change?
- **Possible change:** What do we need in order to get things going as effectively as possible?
- **The desired sustainability:** What will help us make improvement a natural and integral component of our work?

In order to keep moving forward with her assumptions, Ms. Schweizer derives some action points from them: short one-on-one conversations with two skeptics, longer discussions with everyone she suspects of harboring strong resistance, as well as a well-prepared team talk in which she discusses a group kickoff for the change initiative.

"Comprehensive objection removal" is her slogan, which resonates with her experience that "a massive investment in communication is particularly important right at the beginning of a change project."

16.3 THE CHANGE DIALOG

What should you do when you are completely convinced that Kanban is the right change option but some team members are nevertheless somewhat skeptical? What if improvement ideas are met with reservations right from the beginning? And what if your initiative seems to draw resistance to the plan rather than commitment? To answer these questions, let us return to Ms. Schweizer. In her team constellation, the experienced Head of Development had identified three subgroups as well as one clear outsider, to all of whom she had assigned a specific mindset with regard to her Kanban initiative:

- An **open attitude** to the initiative, that is, curious, engaged, and already committed to continuous improvement
- A fundamentally open-minded but nevertheless partly **skeptical attitude** toward Kanban
- A **reserved, even dismissive, attitude** that sees little or no good in Ms. Schweizer's initiative
- An **outlier** whose presumed resistant attitude is already giving Ms. Schweizer a headache

It was then time to get in touch with the individual team members and risk a comparison between Ms. Schweizer's conclusions and their actual perceptions. Due to her previous experience with the culture of the family-run organization where most of the participants had been working for many years, Ms. Schweizer anticipated a headwind. However, due to her understanding of change, it was clear to her that working with these objections was very important: interaction with what is generally referred to as "resistance" is crucial for the greatest possible success.

EXCURSE: RESISTANCE AND CHANGE

"Wherever there's change, there's resistance." So goes the saying. But what does it mean?

- **Resistance and change are Siamese twins**. Despite our negatively influenced, everyday understanding of it, resistance doesn't actually have anything to do with "evil intentions" per se. Systemically, resistance refers to a force opposing change that is oriented toward preservation. While one force tries to set itself in motion, the other force holds it back.

- **Resistance can be caused by a huge range of factors**. Lack of knowledge, absence of interaction with the causes of change, methodological implications, bad experiences with previous change initiatives, personal comfort, and specific departmental conflicts of interests. Resistance is thus neither a purely emotional nor a purely cognitive affair. It is rather a complex mixture of various phenomena that are for the most part subconscious.
- **Resistance appears in all sorts of different forms**. Indirectly: a drop in work energy, a blocked flow of information, sloppy task execution, or bad atmosphere. Directly: increases in sick leave and resignations, the appearance of violent conflicts, criticism of the management, or even official strike action. The drop in productivity represented in Figure 16.2, the so-called J curve, is a typical form of systemic resistance in change projects.
- **Resistance can appear in any phase of change**. Resistance cannot be eliminated once and for all. It is constantly capable of finding new sources of nourishment and can appear in the preparation phase of the Kanban introduction just as easily as in the creation of the system design or in operations. Resistance is, as it were, the zombie that continuously accompanies the change monster [2].
- **In practically all cases, resistance is unpleasant**. This is particularly true for committed change agents. Resistance persists, costs time, wears down your nerves, and sometimes leads to physical conflict. It's no different in Kanban. However, experience shows that the earlier resistance is acknowledged and considered a part of professional change management, the greater are the chances of success.

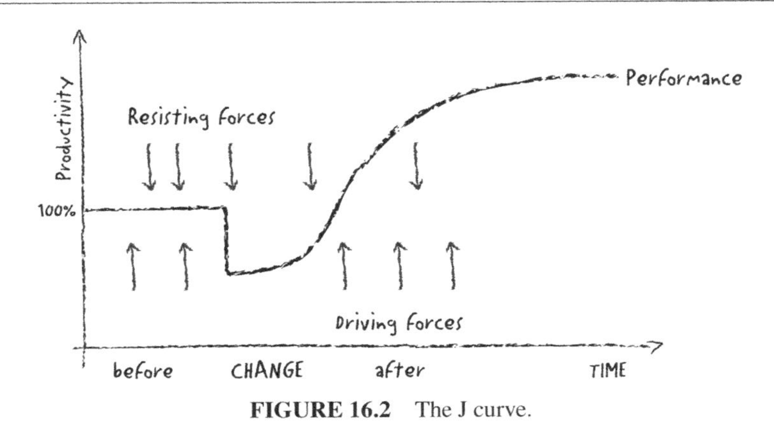

FIGURE 16.2 The J curve.

Ms. Schweizer knows that resistance, defense, and protection are not independent phenomena. All sorts of change impulses are fought off: new demands, conflicts in the team, or the need to question oneself. The protective function of resistance, for example, is evident in the refusal of change, which allows more detailed thinking through things and slows down careless action taking. Resistance offers protection from the hardships of change and creates the time required for all participants to achieve better understanding. And resistance has the positive effect of carefully weighing up what is currently in place against the proposed changes, leading to a better understanding of what should definitely be kept. Change expert Robert Fritz called this push-me–pull-you between change and retention "structural tension" and pointed out that change can only be successfully implemented if this tension is released in a constructive manner [3].

> *Consequently, resistance must not be negated or, to employ a classic of management wisdom, conquered. Instead, it is necessary to access the power locked up in resistance and harness it for the purposes of change.*

How can this be done? Short answer: through professionally modeled communication processes. Longer answer: through the desire to inform those affected by your project as early as possible, the willingness to involve those affected in your project in a suitable manner, and the ability to establish this participation on the basis of broad consensus. The case study of Susanne Schweizer shows what this can look like in practical terms. The goals of her change dialog are as follows:

- A **better understanding of the people involved**
- **Open and frank interaction** with the various attitudes concerning Kanban
- **Personal information** concerning Kanban change management, the desired use, the steps planned, and the contributions desired
- **Setting** the potential resistances to Kanban **in motion**
- **Agreement on follow-up measures**

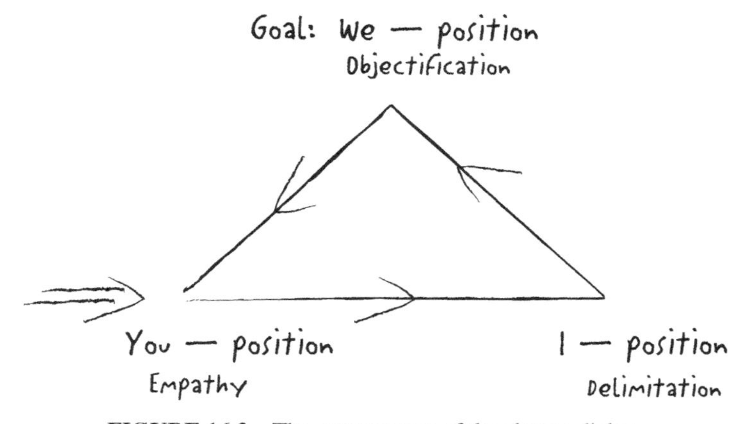

FIGURE 16.3 The cornerstones of the change dialog.

Austrian psychologist Paul Lahninger has created a simple triangular framework for the modeling of the change dialog [4]. Figure 16.3 outlines this model using the three cornerstones: "empathy," "delimitation," and "objectification."

16.3.1 Empathy

The first cornerstone of Lahninger's triangle focuses on the *you*. It's about the questions, remarks, and moods of the person sitting or standing across from me. Drawing on the basic dialog principle of respect, the most important thing here is attentive listening. We must accommodate what Rudi Kastner, Petra Freilinger, and Jakob Zwazik really think about the change intention. To this end, Ms. Schweizer uses open questions such as:

- "Rudi, you know that I value you as a critical voice in this team and that I've often profited from your experience. I'd be very interested in your opinion concerning the current situation and what you think of Kanban generally?"
- "Petra, as I said in the introduction, it's very important for me that the change is implemented by everyone together. I can well imagine that not all reactions are pleasant ones. What do you think of the present situation?"
- "Jakob, it seems to me that you don't think much of Kanban. I'd like to learn more about what really gives you a headache."

Ms. Schweizer knows that the *you* phase is primarily about active listening. What you hear must be repeated in your own words, you need to ask how something was meant, and you must explore the other's perspectives and at the same time show understanding for their emotions. A subgoal of this stage is to really learn the actual attitudes of each of the conversation partners. We are in no way concerned with discussion or "correction" here, even though this isn't easy with statements such as:

- Rudi Kastner: "It's nice to know that you value me as a critical voice. That makes it easier for me to voice my discomfort with Kanban. I feel like it's really just about even stricter project management and that you just want to improve the waterfall with hundreds of new metrics."
- Petra Freilinger: "Well, to be honest, it's already brewing a bit. I have the feeling that Rudi is very skeptical. And Jakob's just doing his own thing, same as ever."
- Jakob Zwazik: "I don't quite know what you mean. I'm just going to let Kanban come to me. Then we'll see soon enough."

Whether Susanne Schweizer is now faced with criticism of the methodology, worries about the team spirit, or rejection, as far as the professional dialog is concerned, it's crucial that she keeps her own reactions in check during this phase. One's own opinions must be put on hold; the first step of the change dialog is empathy and not confrontation.

We won't win over anyone by constantly commenting on every critical question, every methodological reservation, and every fear voiced. Ultimately, we don't want to conquer; we want to convince.

16.3.2 Delimitation

The exploration of differences, however, does not mean that Ms. Schweizer's perspective is ignored. But the personal background, information, and feedback of Ms. Schweizer are only brought into play in the second stage of the change dialog. Having focused on the *you*, we now move on to the *I*:

- What made Ms. Schweizer think of introducing Kanban?
- What has she promised herself with it? What benefits does she see for the individual employees? What about the customers?
- How would she like to proceed? What will she build on? Where does she anticipate barriers?
- What would Ms. Schweizer like from her colleagues in order to get rid of these barriers?

In conversational terms, professional feedback and clear statements are deployed during this phase. Explicit wishes and demands can also be mentioned. With the delimitation step of the change dialog, the need for change is fleshed out. In this respect, delimitation means not only correcting false assumptions ("Kanban is just a more elaborate waterfall") but also reacting to concerns ("Kanban will tear apart our team's network"). The delimitation of open passivity is just as important ("I'm just going to let Kanban come to me.")

16.3.3 Objectification

The main goal of the change dialog is the release of tension. This means that the third stage of our change dialog is bringing together the *you* and *I* perspectives as well as possible—not only at the technical but also at the emotional level. To this end, Ms. Schweizer employs the following tools: a summary of the most important arguments, the establishment of conclusions, and the agreement on real, concrete follow-up steps. So, for example, Ms. Schweizer could say:

- "Rudi, if I may just quickly conclude by summing up our conversation, it was particularly important for you that Kanban is not implemented top-down. Everyone in the team should be involved and the available strengths should be well implemented. I have also gathered that you are unsure as to whether the stakeholders will really adhere to the agreed WiP limits. Is that correct?"

- "I'm relieved to hear that I've understood you correctly, Petra. For me, it's just as important to shape the change in a team-oriented way. We will be availing ourselves of external support in the preparation stage. I hope you've understood that our collaboration is the most important thing for me. May I rely on your continued critical but fundamentally open-minded support?
- "Wonderful, thank you. Now before we get started developing our Kanban board, I'd like the next step to be a group conversation concerning our intentions. I would also like to summarize the most important issues from the individual conversations. Jakob, I hope that this is agreeable to you?"

THE CHANGE DIALOG

- Framework
  - Personal invitation: Why is there a dialog? How will it proceed? When? Where?
  - Quiet location for intimate, one-on-one conversation.
  - Sufficient time (a minimum of 30 min).
- Opening
  - Thank you for your time!
  - Outline the framework, goals, and process once more.
  - Emphasize the importance of the conversation.
- Empathy
  - Ask simple, open questions. Explore differences in opinion.
  - Put your own opinions on hold; stay in control of your emotional pattern of reactions.
  - Promote comprehension via paraphrasing and summing up.
- Delimitation
  - Provide illustrating information, clarify contexts, and correct wrong assumptions.
  - Express your own point of view.
  - Promote comprehension via targeted feedback.
- Objectification
  - *I* and *you*; integrate the technical and the emotional; create the *we* perspective.
  - Sum up results and define follow-up steps.
  - Thank you for the conversation!

In all the interventions described here, we're concerned with a goal-oriented accumulation of energies. We want to build bridges on both the physical and the personal levels. We want to achieve what the Austrians call "talking ourselves into agreement" in order to proceed with the clearest possible understanding.

For Kanban purposes, it is not at all necessary that all tension is released by the end of the change dialog. That would be asking too much. Instead of hyping up the change dialog as a miracle cure, it should be viewed more modestly as a step on the path to continuous improvement and an elementary building block in the creation of a Kaizen culture.

16.4 THE TEAM CONVERSATION

It's normally not the case that all resistance can be converted into goodwill in the space of one single conversation. Indeed, as one swallow does not a summer make, one communication does not an improvement make. Experience shows that a series of change dialogs can affect far more than just a better flow of information: with professional assistance, it can drastically improve the quality of communication, even—indeed, especially—in critical change phases. Particularly at the beginning of the change initiative, misunderstandings can be resolved, different points of view can be aligned, and set-in-stone positions can be softened.

In the context of general organizational communication, change dialogs such as Ms. Schweizer's can be viewed as special cases of the classic staff appraisal. While this staff conversation usually takes place independently of any concrete event (usually once a year), the change dialog follows the sense of urgency typical of change [5]. As a vital starting point for every strong Kanban initiative, this emphasizes the urgency of communication, the basic requirement for any successful change.

In order to drive this improvement forward, our next move is to follow the individual conversations with an integrating team conversation.

Here again, the change dialog is very much in the tradition of the staff appraisal—as we all know, this is always rounded off with a team conversation. What's important for this kind of conversation? And how does it help us get Kanban up and running as well as possible?

In order to answer these questions as succinctly as possible, we'll pursue our case study further.

At the end of each individual dialog, Ms. Schweizer had sensibly already mentioned her intention to speak to the entire team once again upon completion of the individual conversations. This statement already establishes the next stage of the continuous improvement initiative that our Head of Development is striving for with her "Kanbanization." Just as with the team conversation at the end of the individual annual staff appraisals, we are here concerned with summing up the most important themes, requests, and questions distilled from the individual conversations. We also need to make space for our change management. The items "continue to improve the quality of our communication" and "create a strong basis for a successful introduction of Kanban" are on the agenda for this meeting, which Ms. Schweizer has already calmly but intentionally referred to as a "team conversation." We will sum up how Susanne Schweizer converts this conversation into actions.

THE TEAM CONVERSATION

- Preparation: focusing on the goals
  - Summing up the most important themes, requests, and questions from the individual conversations
  - Making change and change management the topic of conversation
  - Improvement of quality of communication between people
  - Strengthening the entire team as a basis for Kanban
- Marking out the boundaries
  - Personal invitations at the end of the change dialog (separate invitations for all those who didn't take part in the dialog)
  - Workspace with a good atmosphere and professional equipment
  - Sufficient time (a minimum of 2 h)
- Explaining the procedure
  - Thank participants for their time and the willingness for an open discussion.
  - Outline the framework, goals, and procedure once more.
  - Emphasize the significance of the workshop: "The reason this meeting is particularly important is…."
  - Bring together the most important themes, requests, and questions from the individual conversations.
  - Ask for brief feedback: What do they notice? What should we occupy ourselves with to a greater extent?
  - Prioritize, for example, using the evaluation of points to weigh up the themes.
  - Process the prioritized themes in small, well-mixed groups with the goal of delivering real suggestions for improvement.
  - Present and discuss the results.
  - Agree on improvement steps: What should we do next? How do we want to proceed? What do we want to concentrate on?

FIGURE 16.4 Conversation results from the change dialogs.

After Susanne Schweizer has once more outlined the framework and emphasized why this workshop is important, it's down to business. Ms. Schweizer presents the group with the most important topics from the individual conversations. Figure 16.4 describes the most significant themes in the change dialogs. These are then later expanded to include the most important observations of those team members who didn't participate in the dialog.

After the presentation, Ms. Schweizer asks for brief feedback: she asks two of her neighbors to put their heads together and spend 5 min going over what they've just heard. "What have you noticed? What are your strongest impressions? Which questions or topics do we absolutely need to discuss?" These questions are written on the flipchart Ms. Schweizer has used to describe the assignment more precisely. This feedback loop has several goals:

1. **Communication of the most important issues** from the change dialogs. For purposes of comprehension, this takes place in natural conversation as close as possible to actual dialog.

2. "Warm-up" between each other. This is the lead-up to the intensive type of communication in the change dialogs Ms. Schweizer has led and now wants to introduce to the whole team.
3. **Collection of feedback** on the presented issues. In Susanne Schweizer's words, "I want to know how high my colleagues raise their eyebrows here." Most important in this phase is the general atmosphere.
4. An initial **weighing up of the themes**.

This is followed by the prioritization of the list of themes covisualized by Ms. Schweizer: two points per person in order to highlight what should be worked on in more detail in this workshop. It's then time to build focus groups. As an experienced facilitator, Ms. Schweizer knows the value of heterogeneous groups; therefore, she makes sure to create a mix of "old" and "new" and "Kanban fans" and "skeptics" in the groups. Ms. Schweizer positions herself as a "jolly jumper" in order to have a presence in all groups and be able to switch according to atmosphere and theme. After a precise definition of the next project—"What can we ourselves do in order to improve our current working situation? Our top three measures are..."—people break off into groups of three. As far as Ms. Schweizer can see, the discussion isn't only very productive but also surprisingly civilized. People let each other talk, listen carefully, and ask meaningful questions. Even controversial issues are discussed in an unusually calm manner.

After the intensive work in small groups, there's the danger that the energy in the plenum drops drastically. In order to keep the attention as high as possible, Ms. Schweizer suggests mixing the groups for the presentation of the results. With one member of each of the previous groups now together in a new group, they progress from poster to poster. The presentation thus remains lively. Indeed, much is already being discussed during this process.

The next phase is especially crucial: agreeing on follow-up steps. To Ms. Schweizer's surprise, however, this doesn't seem to cause the team any difficulties. Several concrete suggestions on how to proceed are made, and Susanne Schweizer jots these down on the flipchart. She is pleased to see that it isn't just those team members she had already gauged as "change-friendly" in her original team formation that are making the suggestions; the skeptics are also interacting positively. And there's even a constructive suggestion from the team's self-appointed "outlaw."

A lightning round (only one sentence each) brings the workshop to a close, which—going by the tone of the lightning round—has not only processed several difficult themes in a solution-oriented manner but has also been conducted in a constructive atmosphere while visibly strengthening the team.

16.5 THE TEAM RETROSPECTIVE

Another method for preparing the team for evolutionary change management is to carry out a team retrospective. If you rate your team's willingness to change highly and don't see any need for clarification conversations, then the team retrospective is just a gentle and effective way for the group to get its bearings. It also offers you as the Kanban initiator a good opportunity to compare the self-evaluation you carried out during the personal retrospectives with the evaluations of the other team members. The actual design of this retrospective is again dependent on the current situation, for example, whether:

- There is already a **culture of regular retrospectives** in the team.
- The team is prepared for an **open discussion** concerning the quality of collaboration.
- You as **the sponsor are convinced** that a team retrospective can effectively contribute to the desired change.
- You have faith in yourself as a **professional facilitator** of this kind of retrospective.

How can the retrospective be designed? To find out, let's look at Herbert Krakauer's situation once more. As the leader of a highly specialized team of experts, Mr. Krakauer is used to tailor-made, individual solutions. What Mr. Krakauer is however less used to is taking time for group reflection on the current events—which only takes place informally, if at all—and he's not at all used to regular retrospectives. However, after a visit to a Kanban basic training program, he's convinced that continuous improvement isn't possible without a targeted analysis of the past. Therefore, Mr. Krakauer decides to get his necessary bearings at the beginning of his change initiative with the help of a retrospective.

Since he himself has no experience facilitating such a retrospective professionally, he begins by enlisting our help. This help is only possible due to Mr. Krakauer's own willingness to recognize the limits of his own experience and to become a role model for intense learning himself. But at the same time, he by no means removes himself from his leadership responsibility, meaning that it is very easy to set the necessary boundaries:

- **Temporal**: 3 h, owing to the team's lack of experience.
- **Spatial**: Large meeting room in the IT department.
- **Technical equipment**: All necessary materials and aids provided.
- **Social**: All team members are personally invited and the retrospective is thoroughly prepared with us.

The team retrospective is opened by Herbert Krakauer, who goes over the agenda once more and explains the function of the external facilitators. Before handing over the baton, Mr. Krakauer adds a few personal words:

> I'm perfectly aware that this is an unusual way of working for you guys. Some of you will perhaps think that it is a waste of time and that it doesn't fit with our concept of Lean IT. Even I was skeptical at the beginning. As you all know, since we're part of a highly dynamic area, we need to be able to move and act more flexibly than ever. I am absolutely convinced that Kanban will provide highly effective support for us here. I

> ask you all to give this mode of change a chance and to treat today's retrospective as a powerful starting point for a new era. Thank you.

The department leader's words seem to have an effect. Initial feedback to the question "What do I expect from today's retrospective?" already provided many positive signals: "We are still skeptical, but we're up for it!" "We expect an open discussion and tangible improvement steps." "It's great that we're finally taking the time to look at these fundamental questions." And particularly noteworthy: "It would be stupid if we were unable to progress towards success in stormy weather!" With this kind of tailwind, it is almost superfluous to insert the prime directive here as a kind of cultural framework.

THE RETROSPECTIVE PRIME DIRECTIVE

> Regardless of what we discover, we understand and truly believe that everyone did the best job they could, given what they knew at the time, their skills and abilities, the resources available, and the situation at hand [6].

Having set the stage and created a certain amount of safety, we then take the next step in Esther Derby's and Diane Larsen's phase concept [7]: the gathering of data. Since Herbert Krakauer had already collected countless data in his do-it-yourself retrospective, we decided to make use of the same network of questions for the team retrospective. Accordingly, participants were invited to individual brainstorming sessions focusing on the following points:

- What's gone well recently? What successes were achieved in my area of work? Which improvements worked?
- What's not gone so well recently? What failures have there been? What's gotten even worse?

In order to get into a deeper discussion as swiftly as possible, a simultaneous presentation was carried out. All team members pinned up their cards on the two available pin boards marked "+" and "–." The facilitator proposed a simple clustering so that the team wouldn't sink into information overload. After a brief group viewing and the clarification of a few questions ("What's meant by xxx?"), the next stage, dedicated to generating insights, began. The following question was discussed in groups of three or four:

- What conclusions can we draw from this balance? What should we absolutely keep in place in future? What do we urgently need to change?

After 30 min, each group presented their top three suggestions for "keep in place" and "change." The good team spirit, the depth of technical knowledge, and the fundamental satisfaction of the customers were all mentioned several times as worth keeping. The main things to be changed were the work overload, the lack of structure for project clearance, the frequent chaos in everyday work, the frequent disruptions, and the latent impatience of sales and distribution.

Possible improvement measures for these suggestions were discussed. In this fourth and final stage of the retrospective, Mr. Krakauer once again took on his usual leadership role. He praised the results, outlined the course to be followed in terms of the Kanbanization, and expressed his pleasure for the success of this kickoff in which pleasure was shared by many in the final round. Much was said of the "positive surprise" that "we've finally started talking simply and clearly" and that "together we've brought the most important problems to light." Of course, there were also people who wanted to remind the team that "although this is of course an important first step, we've by no means already achieved our aims" and "whether we really achieve this is yet to be seen." However, all things considered, Mr. Krakauer can say that his experiment has had very positive results at the team level as well as in terms of real approaches for methodological improvements.

THE TEAM RETROSPECTIVE

- Focusing on goals
  - Joint reflection on events; shared learning
  - Getting bearings and determining outlook
  - Establishing real improvement steps: what can we do ourselves?
- Establishing boundaries
  - Personal invitations
  - Good location, sufficient time, professional equipment
  - Thorough preparation, possibly with external assistance
- Creating safety
  - Define the desired culture—possibly with the prime directive.
  - Personal message: What is important for me? What do I want to contribute?
  - Initiate warm-up dialogs.
- Gathering data
  - Brainstorming with cards (one answer per card) or verbal questioning (facilitator visualizes)
  - Presenting and/or clustering the "+" and "–" answers
- Generating insights
  - What conclusions can we draw? What can we keep? What must we improve?
- Deciding what to do
  - What can we do? What steps do we want to take?
- Rounding off
  - Miniretrospective of the team retrospective: What was good today? What can we try to do better next time?

16.6 THE STAKEHOLDER MAP

It should by now have become clear why we ask you to carefully prepare for your Kanban introduction. Ultimately, it's far from certain that all your employees are convinced of the need for change, let alone are prepared to get involved in it.

And indeed, the same goes for your stakeholders. Together with internal change and resistance dynamics, you also need to concern yourself with the external forces. The last three tools for deeper understanding also help to win your stakeholders over to the idea of evolutionary change with Kanban. In order to be able to mobilize management, customers, and partners for the desired change, you first need to identify them. Indeed, you must go further—you should not only know whose demands and interests are of particular interest for your area of work, you should also know how your stakeholders are connected to each other.

It is necessary to shed some light on the strategic significance of certain stakeholder groups for your value creation, thereby identifying possible coalition as well as conflict partners.

Unfortunately, this is easier said than done, especially when you're bound to have different opinions in your team. Therefore, the map that your team will create needs to achieve the best possible integration of the existing perspectives. Resist the temptation to impose your own map. Instead, allow yourself to be taken on a voyage of discovery. The reward awaiting you is not only a richer exchange of experiences and a better overview for all participants. Since the creation of the map is about step-by-step discovery through constant scrutiny and adaptation, it can also be seen as a significant step toward a culture of continuous improvement.

Duly motivated, Mr. Drechsler sets out to establish the stakeholder map, as planned in the course of the formal team conversation. Knowing that his friend Herbert Riesch had also created a group stakeholder map, he asks him for a written guide.

THE KANBAN TEAM'S STAKEHOLDER MAP

The team stakeholder map is created in five stages:

1. Of course, at the center of your map lies your work group's mission. All organizational units, groups, or individuals relevant for the success of this mission are identified in a brainstorming session and then categorized according to the following questions: Who is particularly important for our value process? And who can be more or less ignored? As facilitator, you can suggest simple dot voting for the evaluation (e.g., everyone determines their top five stakeholders).

2. The selected stakeholders are transferred to cards of various sizes. The size of the card refers to the significance of the stakeholders for long-term success. The goal is a weighting that everyone approves.
3. The stakeholder cards are placed at different distances from the center, that is, your mission. This is to express the extent to which your daily activity but also possible changes affect the various stakeholders. It is important here to experiment. Experience shows that it is worthwhile to gather around a table or at a pin board so that everyone can get involved in talking through the various positions and considering the various points of view. As the facilitator, your task is to limit people's desire to assert themselves and ensure everyone gets involved.
4. The cards are connected to the center by means of lines showing the frequency of contact. A single line refers to little contact, a double line to average, and a triple line to a high frequency of contact. Here again, what's important is the pragmatic understanding that this process can be sped up at any time through the generation of average values (on the basis of the team members' votes for single, double, or triple lines). Experience shows that it is highly productive to *discuss* who in the team has contact with whom and to what extent.
5. Finally, the specific quality of the collaborative work can be made even clearer using symbols, for example, a thunderbolt for a conflictive relationship, a right angle for a blocker, a plus sign (+) for positive, a question mark (?) for an unclear cooperation basis, and so on and so forth. Unanimity is not needed here. Once again, the deciding factor here is open communication within the team ("Why do you see this negatively while I see it positively?" etc.)

Due to his careful preparation, Mr. Drechsler succeeds in leading his team through the work process. After some initial confusion, everyone has been properly engaged and is committed to describing the current situation together in a picture all agree on. Figure 16.5 shows the result of this negotiation process.

In addition to his own work, Mr. Drechsler sees the team map as having multiple benefits:

- The **group creation of the image** provides an overview and orientation for the whole team.
- The intensive communication process itself is **a piece of real Kaizen**. People are constantly developing and rejecting, discussing and disagreeing, and presenting and withdrawing. Things are continuously improving until a point is reached where everyone has a strong sense of agreement.
- The intensity of the work is accompanied with a strong team-building effect. Mr. Drechsler's active facilitation encourages a communication process in which everyone gets a turn to speak and be heard. Through the exchange of the individual stakeholders' personal experiences, the team members not only gain new information but also get to know each other better.

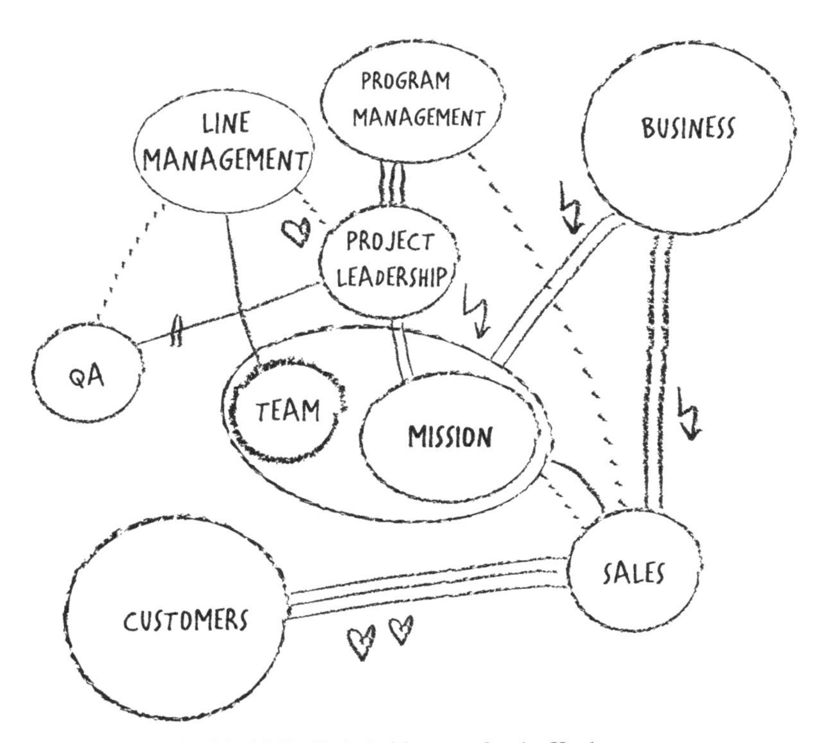

FIGURE 16.5 Stakeholder map for the Kanban team.

- The cautious discussion provides important **stimuli for Mr. Drechsler's leadership**. Firstly, he gets to know his team better. Secondly, the shared map helps to identify some of his blind spots.
- The **improvement work** is brought forward with a great deal of energy. Thanks to the map, the team is on fire. Now, we need to turn this Kaizen fire in the direction of the stakeholders.

16.7 THE STAKEHOLDER INTERVIEW

As we've explained, the stakeholder map helps you position yourself with respect to your most important partners in terms of the improvement work. The logical follow-up is personal contact with these partners.

What are we trying to achieve with this? Let us apply this question to the case study in the preceding text. Sensibly, Mr. Drechsler had already established the follow-up step before the conclusion of the map session: a series of interviews with the most important stakeholders as well as an invitation to a group Kanban basic training course. Everyone agreed that Mr. Drechsler should carry out the interviews himself and invite a Kanban expert for the first course session. All also agreed on a feedback session with the team and all interview partners, during which the most important results would be summarized and further progress would be decided upon.

In terms of the series of interviews, it is particularly important for Mr. Drechsler that each stakeholder that he has personally contacted feels respected and valued for their strategic importance. From his own experience, he knows that he must not underestimate this implicit, indirect message. In many conversations, he expands the implicit message to include explicit praise that he brings in right at the beginning of the conversation. Here are two examples:

1. "Heinz, thanks for taking the time to be here for this conversation. You know that I've valued you as a reliable partner for many years. Therefore, it's really important to me that we discuss your needs in more detail now, at the beginning of our change initiative."
2. "It's great that you accepted my invitation to this brief interview, Marie. I know that you've always had an open ear for our concerns and have often provided helpful tips. Nevertheless, I'd like to put our good teamwork on trial today."

The stakeholder interview is a new opportunity to get together and talk. It helps strengthen contact that is normally dominated by operational pressure and offers both interpersonal benefits and valid indications for further improvement.

The word "interview" emphasizes Josef Drechsler's intention to primarily *listen* during these conversations. In line with the deeper analysis he's striving for, he wants to learn more about what is particularly important to his business partners. How does each of his business partners define success? And what does he or she expect from Mr. Drechsler's team? What already fits here? And where does he or she see the greatest need for change? Mr. Drechsler knows that he must resist the temptation to get into a discussion straight away once the question's been answered. At the moment, it's about better understanding the current situation from the perspectives of each of his stakeholders rather than providing a commentary on them.

Although gathering information lies at the heart of this process, the conversation also offers an opportunity to provide specific information, for example: at the beginning, the goals and purpose of the interview are outlined once more; during the conversation, the needs for change are identified; and at the end, the next steps are explained. With each interview partner and the individual course each conversation takes, Mr. Drechsler takes various opportunities to mention his improvement initiative and above all the reasons for continuous improvement. Additionally, after a conversation with his CTO, he knows he can invite all stakeholders to a Kanban basic training course. Following is an overview of the most important points in the stakeholder interview.

An accurate summary is most crucial for further use of the interview results. After clarification of the methodology, the most important statements should definitely be recorded in written form. It is important to make sure that positive comments as well as criticisms are recorded. The easiest way to do this is by writing every important aspect on a separate Post-it and placing them between the conversation partners. Key

THE KANBAN STAKEHOLDER INTERVIEW

- Preparing the interview
  - Personal invitations.
  - Focus on the goals and purpose.
  - Establish a date for the conversation.
- Setting the stage
  - Thank people for their willingness to talk.
  - Mention the goals and purpose once more, as well as how the conversation has been planned.
  - Clarify any open questions.
- Asking the core questions
  - What is particularly important to you in your work? How do you ensure success?
  - What do you need from us for this? How can we contribute to your success? What do you expect?
  - What already works here?
  - Where do you see the greatest need for change?
- Summarizing the answers
  - What are the most important aspects of our collaboration on which we can build further?
  - Where must we absolutely improve?
  - Visualize on Post-its or small cards and put them in order of importance.
  - Check the results once more and reorder them if necessary.
- Rounding up
  - Provide information on how things will proceed: What will happen with the cards? How do you inform about the results? What further steps do we want to take? How do you benefit from this?
  - Clarify any still open questions.
  - Thank the partner for the conversation.

words and short sentences, as well as short quotes, are tried and tested ways of providing interesting material for the ongoing work.

Speaking of "ongoing work," you should be sure to mention in the interview invitation that you would like to use the results of the conversation at a later stage. It is best to renew the request at the beginning of the conversation so that your partners give you explicit permission to use their statements. We also encourage you to order the improvement cards according to importance. To help assign this order, you can, for example, use the so-called magic genie question: "If I were a good genie and were to grant you your first three improvement wishes, are these the ones you would want me to help with?" If your answer is yes, then you've got your results sorted. If not, you should rethink the weighting.

The conversation is concluded with an overview that is as detailed and accurate as possible. Mr. Drechsler ends this with a double invitation to a basic training course on methodology and to a small workshop in which the interview results are discussed and compiled in a group improvement catalog. You can of course pursue a different course of action. The important thing is to keep the ball rolling and to assure your stakeholders that the interview was an important, but by no means final, change step.

16.8 STAKEHOLDER WORKSHOP

"No action without prior analysis." This is one of the core messages of German change management pioneers Klaus Doppler and Christoph Lauterburg [8]. We take it a step further: "No analysis without feedback." What feedback do we mean here? What purpose does the stakeholder workshop serve in a Kanban introduction? And how can it help us achieve our goals?

The stakeholder workshop is basically about creating a feedback loop with all those who participated in the interviews.

The most important results of the individual conversations are brought together, clarified, and consolidated bit by bit. The various stakeholder perspectives are linked up with each other, and a shared understanding of the current situation is extracted. This is important for many reasons:

- Your most important stakeholders go from **being affected** to **being involved**. Just like in the collection phase, we are here interested in actively addressing and integrating all those who stand to profit from Kanban.
- Apart from the win–win rhetoric, you increase the chances that all participants take away something from the change. By making the current wishes of your customers, partners, or superiors the starting point of your improvement initiative, you show them respect and strengthen their trust.
- The strengths and weaknesses of your current work process are revealed. Thus, different points of view can be more easily explored, information that is lacking can be exchanged, and diverging interests can be compared and contrasted. You gradually strengthen the basis of your initiative.
- Meanwhile, it is not only points of view and interests that are brought together in the feedback. Networking of the stakeholders is also an important goal. Having your stakeholders get to know each other better and communicate better with each other are small but important Kaizen steps: they build trust and have a positive influence on the value process. The workshop also provides a good training opportunity for further networking and group decision-making that will become necessary in the course of the queue replenishment meetings.

Our experience shows that these goals can be achieved in two different ways: either through feedback that is strongly structured around the input of the initiator or through feedback that is by and large created by the stakeholders themselves. The progress of two of our case studies should clarify the differences between these two options.

16.8.1 Feedback Through the Kanban Sponsor

Let us start by taking a look at the case of Roswitha Rösner, who we already know as the Head of Development of a media organization. Thanks to the retrospective in her last change project, Ms. Rösner has also identified her most important stakeholders. In order to better determine her bearings, she has carried out a series of individual conversations as well as a conversation with her entire development team. As agreed, she sends a formal invitation to all her conversation partners in which she again outlines the goal and purpose of the workshop, asking people to note the date. She next goes through the material that she had gathered in the course of the conversations, more precisely the Post-its on which she had noted the most important points from each conversation. These notes are the starting point for the presentation in which Ms. Rösner provides a comprehensive overview of the most important conversational topics right at the start of the workshop.

She has chosen to give a presentation because the participants in the workshop are relatively green when it comes to Kanban. Naturally, Ms. Rösner had informed them all about the fundamentals of her change initiative during the interviews. However, apart from one of her developers, none of the stakeholders had participated in a preparatory training course before. Therefore, Ms. Rösner has two goals: firstly, she wants to use the workshop to reach a common understanding of the problems; and secondly, she wants to convey further knowledge of Kanban by establishing a connection to the problems and ideas from the interviews during the presentation. So how precisely does she go about this?

- During her **preparation**, she once again reviews the answers that were given in the interviews with the team, superiors, customers, other departments, and management colleagues. Ms. Rösner soon notices that certain topics keep arising and also defines some topic clusters. She finds this much easier since she had asked her conversation partners to provide not only their perspectives and motivations but also their respective priorities. The clusters therefore do not gather together all answers but rather those already ordered by the interview partners. These are next categorized: "Which improvement areas can be addressed with which Kanban element?" Table 16.1 is a sample extract from the presentation Ms. Rösner has prepared.
- The **presentation** relies on the interplay between summary and input: firstly, the most important topic clusters from the conversations are presented, and secondly, the group is informed about the Kanban approach. This is as much an overview of the positive things ("What fits well?") as it is about the challenges ("Where do I see a need for change?").

TABLE 16.1 Presentation of the interview Results

Wishes	Kanban element
• "I want to know who is working on what and what problems there are" • "I want to know what needs to be worked on in the near future" • "We want to have an overview of the amount of work there is to do"	• Three classic cases that can be addressed with the visualization element
• "I don't want to have to do a million things at once" • "Projects should be completed properly" • "We want to receive quicker feedback from the testers"	• The WiP limits can help us satisfy these demands

- A small but significant component of the presentation is the use of **the original statements**. In some cases, the interviewees have even recorded their most important topics in such a way that their handwriting is identifiable. Although it is not important in terms of the feedback to know precisely who provided which topic, working with the original text and handwriting has been shown to have a positive effect. The stakeholders recognize their requests within the individual clusters and consequently feel respected and understood. If Ms. Rösner were to change the gathered material during preparation, reformatting or even digitizing them, this could easily be seen as manipulation. Experience shows that such reworking alienates the people involved and creates a token distance between them and the change project.
- Once Ms. Rösner has presented the conversation results step-by-step and illustrated the related Kanban principles and elements, she opens the **discussion**. In order to generate as much energy as possible for this and make sure every voice is heard, she first invites everyone to participate in a so-called **buzz group**. This is where groups of two or three neighbors put their heads together and spontaneously exchange their initial impressions. Ms. Rösner has written the following on a sign in order to focus the murmur groups: "What are the most important results, in my opinion? What do I think of the connection between Kanban and the wishes? Please provide one comment and one question."
- At the end of 10min of murmuring, Ms. Rösner collates the **comments and questions** from all groups on a flipchart. She first makes sure that all comments have been accurately recorded and that everyone has understood them. By reading everything out again, the various statements are evaluated and their content connected: What is mentioned repeatedly? What seems to have made a particularly good impression? And what does this mean for Kanban? Ms. Rösner then goes through the questions, also trying to connect similar gray areas and provide comprehensive answers. A lively discussion evolves, generating further questions based on Ms. Rösner's answers.

At the end of the discussion, two goals have been achieved: the participants have communicated their individual results to such an extent that a **common understanding of**

the most important strengths and areas of improvement in software development has evolved; and Ms. Rösner has been able to provide a little more information about what continuous improvement means and how Kanban contributes to solutions to the identified problem areas. Many questions were answered, misunderstandings cleared up, and several fears—for the time being, at least—allayed.

The workshop ends with an **overview of the next steps**. With the help of an external Kanban expert, the team will develop a so-called system design for improvement. This design will be presented to all the stakeholders participating in the workshop in order to get feedback, make any necessary corrections, and, after all participants' final assent, get the Kanban system up and running. Everyone present is also invited to the Kanban change training course that will take place before the system design in order to provide the required know-how.

16.8.2 Feedback from the Stakeholders

Let's see how Mr. Drechsler's Kanban story is progressing. As the Head of IT of an infrastructure organization, he too carries out a series of stakeholder interviews in order to determine the current situation as accurately as possible. In contrast to Roswitha Rösner, Josef Drechsler even has individual conversations with all members of his team and organizes a 1-day "crash course" for all the important stakeholders with us as external Kanban change management experts. The know-how created allows Mr. Drechsler to model the workshop—which takes place just a few days after our crash course—considerably differently than Ms. Rösner. Admittedly, it still consists of a compilation of the conversation results, but additionally, Mr. Drechsler also uses the feedback for further methodological inquiry. Here is a description of his approach:

- After a brief introduction, Mr. Drechsler presents a **collection of the most important conversational topics** on a pin board, only roughly ordered. Rather than methodological ordering, he presents the participants with two other pin boards, each divided into two columns. The left column of the first pin board is headed "Major strengths: We can build on these," while that of the second pin board is headed "Most urgent need for change: We need to address this soon." Both right columns are headed: "The next steps: How Kanban helps."
- Mr. Drechsler now asks all stakeholders to come to the three pin boards at the front. They are separated into two well-mixed groups and tasked with **creating a sensible ordering together**. The topic cards should be distributed across the two pin boards, clustered, and then connected to Kanban elements. Table 16.2 shows an extract of the groups' results.

The tactile element is particularly important, since things are very literally *hands on* here. Since all participants in the stakeholder workshop have previously completed a Kanban training course, they are already familiar with the most important elements. In this process of **repetition and learning**, we're no longer using fictional examples. We're now **focusing on the actual workflow in the organization itself**. This leads

TABLE 16.2 Presentation of the Group Results

Most urgent need for change: We need to address this soon	The next steps: How Kanban helps
• "I want to know when my work is completed" • "I want due date performance to improve"	• We can achieve this with SLAs, WiP limits, and the queue replenishment meeting
• "I want to know what the team's working on" • "I want to know who is working on my topic"	• We can easily provide this information with the visualization

to interesting discussions in which various points of view are voiced. "How do you guys see it?" Mr. Drechsler says, repeatedly picking at a topic. "Is that the same as...?" "Or is it something else?" "And how could it be improved?"

Due to the **high level of energy** and the **productive self-organization**, Mr. Drechsler stays in the background. Every now and again, he contributes to specific clarifications. Instead of getting too heavily involved and running the risk of disrupting rather than helping, he retires to the position of an observer. What he hears and sees is exactly what he has hoped for. Everyone is getting involved; nobody is standing aside, letting the others do the work. People talk with each other, at times even as an entire group together. The required agreement on WiP limits is raised, as are service classes and meetings. And the importance of the visualization can be felt in the room.

Naturally, Kanban doesn't have a simple solution in its kit for every single problem. Often, there are **many different solutions** for one and the same problem. A wish for "more space for operational issues" can be addressed either with visualization, WiP limits, capacity distribution, or service classes. Working through the various solutions will make it clear that there is no such thing as a perfect Kanban system. Rather, it is the changeability of the system that makes it perfect—including the fact that actual change lies in the hands of those affected. "If something is no longer serving its purpose," Mr. Drechsler observes, "we will rethink it. But first we need a starting point."

In the subsequent **presentation of the group results**, Mr. Drechsler makes a plea to the stakeholders not to let uncertainty due to the many possibilities gain the upper hand. "In the course of this workshop, we are concerned with getting an idea of what the most important problems are and what a solution could look like," he explains. "The precise development of this solution is then the responsibility of the development team designing the system."

Although content is significant, it is also important that all stakeholders present their opinions and discuss possible solutions and approaches together. Ultimately, it's about one thing alone: **communication**. And communication will improve massively as a result of the workshop.

Thanks to the intensive networking, the only thing Mr. Drechsler has yet to do is give an **encouraging summary**. He praises the openness that has gone beyond specialist and hierarchical boundaries and underlines how important this openness is for

THE STAKEHOLDER WORKSHOP

We here summarize what you as a Kanban initiator should watch out for during the feedback workshop:

- Preparing the workshop
  - Send personal invitations to everyone who participated in the interviews.
  - Stipulate goals and purpose.
  - Clarify organizational aspects (when, how long, where, etc.).
- Setting the stage
  - Thank everyone for their involvement.
  - Briefly outline the goals and purpose as well as the planned process once more.
  - Deal with any open questions.
- Focusing on the conversation results
  - Either present and clarify the clusters or illustrate the top results and invite the participants to define and clarify the clusters.
  - Take care of any additions and corrections.
- Making problem areas and solutions visible
  - Summarize the most important wishes and clarify how these can be addressed by Kanban.
  - Plan a suitable visualization (e.g., a metaplan or table).
- Rounding up
  - Inform participants of what comes next.
  - Deal with any open questions.
  - Thank everyone for their commitment and enthusiasm.

future negotiation within the replenishment meetings. He then rounds off the workshop with a brief preview of the further steps as well as a final lightning round on the day's workshop.

16.9 SOLO, DIALOG, COACHING, OR TRAINING?

Allow us a few remarks before we summarize the preparation steps we favor. Of course, we don't want to condemn you to heroic solos during the general clarification and deeper analysis phases. Just as you're free to choose which tools on offer here to use and in what order, you can also decide whether you want to "do-it-yourself" or not. This decision depends on your current organizational situation, the change pressure you're experiencing, your personal style of learning, and how you feel on that particular day.

In practice, an inner monolog represents the obvious starting point from which you then proceed to focused dialog:

- **With peers from your own organization** with whom you have a trusting relationship.
- **With colleagues from other organizations**, for example, within professional communities.
- **With people who already have practical experience of Kanban**—you can find them, for example, through events organized by the Limited WIP Society.
- **With a professional coach**, who will bring Kanban and change management know-how into the mix.

"How can I know what I think before I hear what I say?" Karl Weick asked in his landmark book Sensemaking in Organizations [9] *summing up the purpose of dialog in a very succinct formula.*

As we hope to have shown you, it is not just abstract thoughts that are expressed in the preparation stages. Hearing and physically picking things up and holding them are just as important as seeing. And together with half-baked ideas, every dialog is often filled with subconscious feelings that, if captured, can also emerge as valuable resources.

Weick's formula is not only valid for good conversations. In a slightly altered form, it also explains the purpose of professional training measures. "How can I know what I am able to do before I try doing it?" With Kanban leadership training, you can get your knowledge into shape, try out selected tools, and gather immediate feedback in the company of like-minded people.

All roads lead to Rome. But before you get there, you should take in a couple more stops. The next one is system design.

WHAT YOU CAN TAKE AWAY FROM THIS CHAPTER

The following table presents a summary of the methods described in this chapter that you can use at the beginning of your Kanban initiative to get your bearings as a group. On the left are the change tools with our view of their principle purposes on the right.

Tool	Purpose
Personal retrospective	• Time for personal reflection • Initially get your bearings • An effective way to beginning with oneself
Team constellation	• Focus on most important change agents, that is, your team members • Check your assumptions on their mindset • Visualize their commitment to the change initiative and to each other

Tool	Purpose
Change dialog	• Personal contribution • Intensive listening and focused information • Reduction of resistance and mobilization of change energy • Commitment to further steps
Team conversation	• Summary of the most important topics from the individual conversations • Feedback and contributions • More specific work on individual problem areas • Real improvement measures
Team retrospective	• Analysis of the starting point of your initiative • Highlighting of strengths and areas of improvement from the team's perspective
Stakeholder map	• Overview of the most important stakeholder groups whose perspectives should be taken into consideration • Alignment between people's own fields of work and the broader environment
Stakeholder interviews	• Personal contact with the strategically important partners in your value chain • Strengthening of collaboration • Winning the stakeholders over to evolutionary change
Stakeholder workshop	• Bringing together the most important stakeholders • Working on the results from the interviews • Creating a shared catalog of demands for the system design
Working with internal or external coaches, facilitators, and peers	• Technical input from other experts • Exchange at the peer-to-peer level • Cocreative collaboration with consultants

17

THE SYSTEM DESIGN WORKSHOP

Finally, after all the intensive preparation critical for the success of a Kanban initiative, we are finally ready to design the team's actual Kanban system using the system design workshop. This is the system design that the team will work with on a daily basis while constantly changing it. We will firm up the points from the first part of this book step-by-step using examples:

- **Identifying the work item types**: Which stakeholders allocate which work items and to whom are they passed on once the Kanban team has processed its work items?
- **Identifying the work steps**: Into what stages are the activities for the individual work item types broken down?
- **Defining the capacities and WiP limits**: How much work can a particular system accept while maintaining a continuous workflow?
- **Determining the classes of service**: How do work items differ in terms of impact and risk? Which services can be delivered and within what time frames?
- **Defining the measurements**: What information about the various functions of the system should be collected in order to glean potential for improvement?
- **Agreeing upon an operational cadence**: How frequently should meetings be held?
- **Concluding the workshop**: Simulation of the entire Kanban system and feedback round.

Kanban Change Leadership: Creating a Culture of Continuous Improvement,
First Edition. Klaus Leopold and Siegfried Kaltenecker.

In practice, you will notice that the system design only functions smoothly if the preparatory stages have been carried out diligently. It is inevitable that questions will appear at this stage whose answers can be traced back to the initial conversations, motives for the change initiative, and interactions with the stakeholders. Armed with these results, the system design becomes clearer and, crucially, the people affected become more cooperative. With the introduction of Kanban, we see, time and time again, how people try to use the classical form of process design. Usually, the starting point is a certain level of frustration: the existing process is not being adhered to and a solution to this problem is therefore sought. But the problem here is that the actual *cause* for the process not working is not sought. Could it perhaps be that the process *itself* is the problem and it's simply unable to cope with the actual demand? Full of excitement at the idea of "just giving it a go with Kanban," the concerted interaction with the participants is ignored, and a supposedly perfect Kanban system is developed by individual process designers, of course with a view to resolving only the current problems rather than creating a process of continuous improvement. The visualization provides pressure in this case: if a process is visible, then the temptation to ignore it is lesser. This is not the case however. The employees don't work with Kanban in a way suitable to their actual situation, but rather, they are forced into a system designed by someone who doesn't need to exist in it. This immediately contravenes two points that we cannot repeat often enough in this book:

1. It contradicts the notion of evolution since we don't develop a system that remains irrevocably valid for the rest of time. Kanban is not a solution that glosses over weaknesses in the process or even cements the existing process, but is rather a way of opening these systems up so that they can be improved. The system must constantly be able to change as new needs arise. This is why it is a contradiction of the notion of evolution when process engineers develop the Kanban system *for* the employees. Moreover, the first principle of Kanban is "start where you are" and not "start where you would like to be." If you are trying to implement a system that has already been designed, you're no longer starting from the status quo.
2. As we have already shown, change is not always welcomed with open arms. If employees—and at the end of the day they're the ones who are going to have to live with it—build the Kanban system themselves, if they establish their own policies, they are less likely to go and torpedo the entire system. They will identify themselves with it because they understand it and they will be interested in improving their own design on an ongoing basis. Kanban is a system of openness not of force. Knowledge work must remain agile in order to remain capable of finding new solutions. In software development, we're dealing with intelligent people who create clear structures out of complex technical material on a day-to-day basis. Therefore, they are also people who are capable of designing the processes they use for this purpose.

Let's use the example of Peter Dunkel, leader of the quality assurance department and process manager of a major financial services provider, to explain this. This is the only appearance he makes in this book, since he had to admit that his Kanban approach was doomed to failure. Why? Because he didn't bother with the preparatory stages and jumped in at the deep end. And the result was: back to the start! So what had happened? For a long time, Peter Dunkel had observed that his employees didn't have a particularly positive attitude toward the prescribed processes. "I wanted to find out why there was such massive resistance to what was in place and find a way of making adhering to processes more palatable for people," Mr. Dunkel said, explaining his intention to us. "In conversations with friends and colleagues, Kanban often came up and I thought the evolutionary approach would be the right solution in my particular case." Right idea, wrong implementation.

Peter Dunkel wanted a new solution, but he used the old methods. Without conversing with his team beforehand, he analyzed the processes in the comfort of his own office, drew up an imposing Kanban board, and fiddled around with the measurements that he wanted to use to control progress. But he'd forgotten something: with Kanban, we measure the performance of the system and not the performance of the individual employees. But this was precisely what Dunkel's suggestions measured: the processing time of tickets per person, the number of blockers on the tickets per person, etc. And he was promptly given a taste of the medicine that he had brewed without the team's involvement. He planned a half-day meeting to share his ideas with the team or, to put it another way, to tell the team where it was at. "I had barely presented the first suggestion when I was absolutely ripped apart," Peter Dunkel complained. "Like a volley of gunfire, I was beaten black and blue with the weaknesses of my system suggestion. In many cases, I simply had to admit the team was right, although some of these were clearly construed special cases that almost never come into play. I hadn't considered some of the issues they have to deal with every day. My Kanban system was a complete flop, but at least it became clearer to me where my employees get stuck in the process and why they are always looking for secret passages."

Mr. Dunkel decided to begin again right at the start, but this time, the team and stakeholders would be included from the word go, because it is only with these people that the picture of the current situation is complete.

We repeatedly experience cases like Mr. Dunkel's: if those affected by a Kanban initiative—particularly the team—are not included in the considerations regarding the design, they will try to bring the system down. They will not perceive it as support for their work but rather as a disguised attempt at control that must be repudiated. Just as processes far removed from the reality of those affected are undermined, so will a Kanban system that is forced on the employees. What results is not a culture of improvement but a culture of information hiding. Therefore, the more sensible path to system design doesn't follow the model of the unfeasibly perfect system (think of a superfast car that consumes no petrol, never breaks down, and also flies) developed by individual people, but rather a **workshop with all those involved**: the people who know best what their work actually involves.

WHO LEADS THE WORKSHOP?

What a Kanban board looks like and the ingredients you need for a technical Kanban system can be explained very quickly. So that means that anyone who's had a bit of experience with Kanban could actually lead the workshop, doesn't it? We strongly advise against proceeding without actual Kanban expertise.

Far more important than the question of whether the experts should be internal or external is the question of competence. An experienced Kanban expert:

- **Offers comprehensive understanding of process techniques and why they work**. In a system design workshop, many complex relationships are brought to light. Through experience, an expert can help understand these relationships and suggest how to deal with them; for example, the point is not to apply WiP limits just for the sake of applying them but rather in order to comprehend their significance.
- **Is a competent facilitator**. System design workshops can last an entire day; some even span 3 days. Therefore, you need someone who can maintain an overview of the events and open questions, lead the reflection upon these events, and steer the group dynamic in a goal-oriented way. Experienced facilitators have developed a knack for sensing whether or not people have really understood something or whether they're just smiling and nodding so that the workshop is over as quickly as possible.

While we introduce the individual phases of this workshop, bear in mind the following:

- **The steps must not necessarily be conceived only in this order**. In practice, the leader of the workshop should arrange a rough guide but also remain attentive and decide on the most productive approach and the topic that needs to be treated more thoroughly in any given situation. Only in the rarest of cases will the individual steps of this workshop be completed entirely. The deeper the team goes into the actual system design, the more frequently will it need to correct findings from previous steps. Thus, involvement with the real workflow leads to the situation where previously identified and grouped work item types must be sorted anew. That's fine: *go with the flow*.
- Apart from being flexible, the workshop leader must have a **clearly defined goal** in mind. The team needs to build a Kanban system with which it can **start**. It's not only process engineers that have a tendency to overengineer: software developers also succumb to this temptation. The workshop facilitator must realize when a team is getting bogged down in details. "Let rip" should be the slogan, rather than patiently working through all scenarios that rarely, if ever, crop up in order to create a completely watertight system that is immune to all possible weaknesses. Sometimes, it is hard to get around the idea that such initial systems last about a week. It's also pretty common that a system is

changed yet again between the design phase and actual implementation. The workshop leader should make all participants aware of the fact that these frequent and quick changes are not bad and that it doesn't mean that a team has carried out bad design work. One point must be made clear right at the beginning: when a system doesn't fit, it is to be changed, even though this might continue happening over and over. The team must be clear about this goal.

- In the course of the workshop, the team will come to an **agreement on the necessary concepts and approaches** for the subsequent continuous operation of the system. Through the time invested in the individual design elements of a Kanban system, the **policies** that the system and its agents will follow evolve more or less as a by-product. Therefore, it is wise to have your own flipchart ready to note down these policies during the workshop. However, be warned against exaggerated precision in defining these policies. Just as with the system design itself, they are not concerned with creating an endless documentation of all eventualities: they are *aides-mémoire* relevant to the majority of events in everyday working processes, because when you create them, you are thinking about the cases that constitute the rule and not the exception. A good starting point for this is a discussion, regardless of whether things are written down or not. If the decision-making process experiences certain difficulties, then the ultimate consensus should be recorded. In all probability, in doing so, the team will hit on a policy. If everyone's clear about everything regarding a certain topic, nothing needs to be written down.

The wish for absolute security is one of the greatest pitfalls in the introduction of any new method, not only Kanban. Policies are a great target for those affected by change and wanting to show that a Kanban system cannot possibly function at all. These people are driven by a fear of the unknown. But 100% security is impossible, particularly in knowledge work. As a complex system, knowledge work is by definition racked with insecurity. The workshop leader thus has the job of steering attention to those policies that address the majority of the everyday goings-on as opposed to the small number of exceptions that rarely occur. With Kanban, we want to promote the exchange of knowledge between employees. If a new member joins a team, they shouldn't have to rummage through piles of written policies in a corner. A new employee should be able to accompany their colleagues, ask questions, and find out what's important for their work through conversation:

- The interviews with all relevant stakeholders, whose results were compiled in the reactions section, were a crucial tool in the diagnosis phase. All those involved in the Kanban initiative had the opportunity to identify problems in the current processes and express their **wishes for improvement**. In the design phase, these results help us build the Kanban system such that it can contribute to solving these problems. During the system design workshop with the team, the interview results should be prominently displayed in the room so that comparison between problems and system solutions is always possible during the workshop. Throughout the process, while referring to the interview results, the team and the workshop leader will constantly ask the question, "Have we addressed this problem sufficiently for the time being?"

- **Make your role clear**. Do you want to participate in the construction of the Kanban system yourself, or are you the nonpartisan facilitator? You can take on both roles, but you must always make clear to the rest of the team the "role" you are playing at any given time. A simple solution: wear a cap when working as a team member in a small group and want to contribute your own point of view. As soon as you revert to the facilitator role, take the cap off again.

This workshop improves the team members' understanding of Kanban and transfers the new know-how to actual current problems. This practical relevance helps protagonists of the change begin to better understand which tools—for example, WiP limits, work item types, or classes of service—would have an impact on the practical implementation of the solution and in what way.

So let's get started with our Kanban system!

17.1 IDENTIFYING THE WORK ITEM TYPES

The compilation of interview topics in the reactions section established the goals to be achieved with the change processes we are embarking on. This final step of the diagnosis phase is also an important element of the practical implementation, not in terms of the development of the system itself but in establishing the direction the change should take and the Kanban instruments to be used—and indeed how they are to be used—to achieve the goals. Following the route defined by these safety fences, so to speak, the team establishes its change route in the system design workshop. Along the way, the team will encounter junctions, multiple-lane stretches, speed limits, tollbooths, and narrow curves with traffic lights. By identifying the work item types in the next step, the Kanban team begins with the actual construction of this route, the actual work of a technical Kanban system.

Head of IT Herbert Krakauer took our advice and in the diagnosis phase generated stakeholder maps himself as well as together with his team, as we recommended in Chapters 15 and 16. These first maps were not primarily concerned with the question of from where the team receives its work items. In the diagnosis phase, the stakeholder maps helped Mr. Krakauer and his team to identify the organizational units, groups, or people that influence—particularly in the strategic sense—their field of work and what the relationship with these stakeholders is: is it overburdened, or is everything working without a hitch? They were thus able to identify the interfaces that needed more intensive communication in order to establish the Kanban change project on a solid foundation.

However, things are even more concrete for Team Krakauer in the system design workshop. It's clear to all involved that at this stage all boundaries have been established and all those affected must support the Kanban initiative. *We therefore now take the identified stakeholders and determine the ones that have direct contact with the Kanban team, indicating that the team receives work items from these stakeholders as input or passes on work items to them as output* (see Section 3.3). In doing so, the team defines the inflows and outflows that work items follow into and out of the Kanban system. Furthermore, through its focus on these communication interfaces, the team also identifies what types of work items the majority of these are. Herbert Krakauer notices that during this process the team members always get sidetracked into discussions about how they normally solve this or that work item, what always winds them up, etc. He therefore steps in, calmly but firmly, and makes things clear once and for all: "At this stage, we are only concerned with casting light on *what* we do and not on *how* we do it." And he shows the team that some questions that actually only come up on the agenda at a later stage are already being answered here: Who must—or should—the team invite to the queue replenishment meeting?

17.1.1 Filling the Stakeholder Map with Work Item Types

The team has now agreed on who gives them their work items and to whom they pass on their work items. But what kinds of work items does the team process? To find this out, let's see which route Herbert Krakauer pursues with his team in the system design workshop.

His team consists of 14 people. With a group of this size, the plenum discussion would be an ideal stage for chatterboxes and a convenient hiding place for those who shy away from talking. Therefore, Mr. Krakauer decides to split the group into subgroups of four. It is important to him that everyone has a chance to speak during the workshop because the change should be supported by all. And everyone should speak *with each other* since experience has shown that this improves the quality of the results. "Please discuss the following question," he says, drawing attention to the next point. "**Which work items do we receive** from our identified stakeholders and **which work items do we pass on** to them?" Shared understanding is the most important goal here. In their small groups, the team members list the work item types they constantly process and present them to the plenum. Herbert Krakauer is quite surprised by the results: a total of 20 work item types are reported. "How can we have reached so many work item types?" he asks himself reflectively. But he can't immediately find an answer. Firstly, his team consists mostly of specialists who receive requests from all sides, and secondly, something else also becomes clear when Mr. Krakauer takes a closer look at the results in the plenum together with his team: in many cases, the same work item types simply had a different name.

This happens in many workshops: the first result is a list of work items that might be short but could equally be quite lengthy. This list is completely ungrouped, redundant, and rich in detail because at this point in time the context of their work is often not yet clear to the team. However, it is immediately clear to everyone in Mr. Krakauer's team that the list of work item types cannot be infinitely long but should instead be reduced to the most important things. In order to subsume the 20 work item types under useful headings or "family names," our head of IT draws once more on the idea of the cluster. A little while later, the team has reduced the number of work item types from 20 to 7. "Seven work item types are actually still rather a lot, but if you want to make a start with them, that's fine by me," Herbert Krakauer says to his team. "In all probability, things will continue to change during operation." And he's right: after the first retrospective, the number of work item types is reduced to five; and after the second, four.

You can thus be sure that the list of work item types you identify in your own system design workshop with your team is only a temporary situation. At the very latest, when things are operational, it will turn out that during the design process, due to the traditional safety reflex, people were still thinking about all the possibilities that actually barely ever happen in real life, although, by the same token, it could actually be the case that new work item types surface. What you're doing here is really only the beginning of a continuous reclustering process that will progress during daily use.

17.1.2 Criteria for Decisions Concerning Clusters

The facilitator now needs to create an overview to help decide whether the number of identified work item types is workable, whether the boundaries between them are clearly defined, and whether they could therefore justify their own swim lanes. If this isn't yet the case, then start clustering. Look for similarities between the individual identified work item types in order to combine them in a sensible way. Before giving the work item type a name, together with the team, take another look at the results of the interview compilation in the feedback. What goals do you want to achieve with your team? What does the grouping of the work item types need to signal so that it is immediately identifiable in terms of how a specific work item should be treated?

In Section 3.3, we provided a few suggestions of how work item types can be grouped together:

- Type
- Origin
- Size
- Rate of arrival of the work item

Although dealing with the work steps of the system process in the workshop usually only comes after the identification of the work item types, similar work processes can of course also be considered a grouping criterion.

CLUSTERING: A TASK FOR THE WHOLE GROUP

With the identification of the work item types, we have seen that clustering shouldn't occur while the participants are split up into small subgroups, but rather together in one big group. By this stage, a fundamental shared understanding of stakeholders and work item types has normally evolved. However, due to concepts and even individual words, it can sometimes take a little longer. In such a situation, having a discussion in smaller subgroups would likely lead to increased fragmentation rather than consensus. The facilitator therefore has the task of clarifying these differences in opinion as much as possible and, with the assistance of the entire team, sorting and grouping the identified work item types thematically.

In actual fact, the next step would be to find appropriate headings or "family names" for the clusters that have been created. Later on in operational flow, these family names will help every team member instantly identify how a ticket they've picked up is to be treated. But before you concern yourself with the family names, look very closely at how this clustering works. Are the cards pushed back and forth frequently? This is often a sign that there is still considerable uncertainty as to whether you can group the work items together in this way or whether they should remain independent because they are their own types. If you notice this uncertainty, let the team spend a bit more time with the naming in the plenum. In such a situation, the team members need to be aware of the broader context in order to be able to better orientate themselves. It is usually the case that the cluster ordering is made even clearer during the search for appropriate family names. Sometimes, it's of course changed around completely.

One question that must be urgently considered is: Should we be writing down a set of policies stating which individual work item types are to be found in the various general groupings? Experience shows that this is only necessary in a minority of cases. The team members have long known what they do the whole time, so the organization of this into a comprehensive thought process is a mere formality. But taking note of agreements when concepts couldn't be split from each other in the preceding discussions or when two different names were used to identify the same work item type can be very useful. We experience this particularly frequently with the three work item types: "Change Request," "Feature," and "Bug." It is subsequently important to note the final shared understanding as a policy itself, for example, "By 'Change Request,' we mean…." Memory is volatile so things that people only agree upon after a long discussion are therefore better recorded on paper in order not to run the risk of being forgotten.

17.1.3 White Noise: Background Voices

For Stefan Bergmüller's (team leader, second-level support) team members, the work concerning the stakeholder map in the diagnosis phase is a form of psychological hygiene. "There's always someone who comes and wants us to have a quick look at something" is the sort of whisper that circulates the team. "Nobody realizes that lots of 'brief' work items add up to a pretty large chunk of time." The stakeholders' uncoordinated wishes constantly make the seven employees dizzy. From Mr. Bergmüller's team's perspective, the Kanban initiative doesn't only have altruistic motives. Of course, the team wants to help the organization as a whole with the improvement. Of course, it wants the customers to be able to rely on the commitments the team makes. But in the course of the preparation, it becomes clear to the team that it sometimes has to be egoistic in order to be altruistic. In Kanban, we also want to signal stakeholders that efficient ways of working require discipline—and this includes the stakeholders themselves. But the fact is that in the course of a working day, teams are often confronted with high levels of the so-called white noise: all the in-between work items that in normal Kanban operation would never make it onto the board because they "only" need a few minutes or hours but are nevertheless just as important. Normally, in the course of the workshop where we identify the work item types, the question arises as to how we are meant to deal with this white noise. Is it in fact an own work item type?

The egoistic goal of every Kanban team is to reduce disruptions as much as possible. It is also one of the most frequently mentioned wishes in the interviews during the preparation phase. Of course, one possibility would be to inform all stakeholders in no uncertain terms that as of tomorrow the team will only be accepting work items that reach them via the established path through the queue replenishment meeting. Apart from outrage, considerable resistance, and questionable hand gestures, this move probably wouldn't achieve a great deal. Mr. Bergmüller's team doesn't want to alienate the stakeholders either. With white noise, the real problem is exactly the same as with all work items in knowledge work: neither can the contractor nor the team see how much this background noise burdens the system and the critical point at which everything else gets blocked up. The best solution therefore is to do exactly the same thing with white noise as we do with everything else: make it visible and thus measureable and comprehensible.

17.1.4 Variation 1

Team Bergmüller uses self-adhesive, removable index markers that you would usually use to mark your place in a book or a document to each represent a 15-min time window. Before this, the team explains to all stakeholders working directly with them what the markers mean. Then, every time a team member receives a white-noise work item, they stick a number of markers on their monitor corresponding to the time the job takes, not just so that they themselves can see them but so that they're easily visible to everyone else too. As the day proceeds, Stefan Bergmüller's team members no longer sit in front of monitors but in front of brightly colored hedgehogs. And they notice two things:

1. At the day's end, the proud owners of the hedgehogs know exactly what they've done all day. By counting the markers, they are able to calculate the amount of time they've spent processing white-noise requests. After work, the team collects the stickers and for every four 15-min markers generates a ticket and places it in the "done" column of the Kanban board. The team also analyzes from whom the majority of these requests originate. The goal of this process of highlighting the white noise is therefore not implementing a new time management system but rather gaining the insight: Where does this background noise come from, and what influence does it have on our work? With this information, a Kanban system can be adapted in order to better process white noise in the future. It might even turn out to be worthwhile inviting those responsible for the noise to the queue replenishment meeting on a regular basis.
2. The team members are also particularly pleased to notice that, all of a sudden, the stakeholders are holding themselves back a lot more. Every potential white-noise originator who enters the team's workspace sees straight away

whether or not the request has a reasonable chance of being processed. Even this simple visual signal with colorful index stickers motivates many people to think a little further about these short work items.

17.1.5 Variation 2

Even if background noise is not officially visualized on the Kanban board but rather on the team members' monitors, you can introduce WiP limits. The team can set itself the policy that each team member is only allowed to process a certain amount of white noise, for example, nobody is allowed to spend more than 2 h a day on such work items. If we retain our 15-min markers, that means eight stickers and that's it. Our visualization supports the rationale with regard to noisemakers too here: the hedgehogs on the monitors and the Kanban board in the background showing that very little has progressed the whole day due to the white noise make it far easier to observe that no further work items can be processed at this time.

Just so there's no misunderstanding: we're not interested in eliminating white noise entirely. As we've already said, these are usually important work items that need to be taken care of. The primary goal is to make clear to everyone involved *what* has an influence on the system, *how* this influence manifests, and to what *extent*. In doing this, we shouldn't be alienating anyone but rather spreading the change generated by Kanban beyond the boundaries of the section in which it's being applied. It's about awareness that there are limits of endurance and that you can achieve your goals quicker with an ordered, sequential processing of work items than when continuously switching between them.

Incidentally, dealing with white noise is a point that should be absolutely incorporated in the team's policy book!

SUMMARY AGENDA POINT 1: IDENTIFYING WORK ITEM TYPES

Marking out the framework

- The results from the interview compilation should be displayed for all to see—they are an indication of the goals to be achieved in the overall workshop.
- Flipcharts are used to record the policies governing concepts and ways of dealing with individual elements of the Kanban system.

The process

1. The team determines all identified stakeholders *from whom* they receive work items and all those *to whom* they pass on their work items.

2. Filling the stakeholder map with work item types:
   a. *What work items* (work item types) does the team receive from the stakeholders and which ones does the team pass on?
   b. Clustering of the agreed work item types into broader groups (e.g., type, origin, size, rate of arrival).
3. Finding options for dealing with white noise. How much of a burden on the team's workshop are "quickie" work items? Two possible highlighting variations:
   a. By physical representation with index markers on the team members' monitors
   b. By introducing WiP limits for white noise per employee

17.2 IDENTIFYING THE PROCESSES

On we proceed, honing our Kanban system. The team now knows from whom it receives work items, to whom it passes on work items, and the work item types for the former. Work item types migrate through various stages of the process, and for this reason, the team must find out what the standard job process looks like. But here again is a typical trap: we are not trying to *define* the process we would like to have; otherwise, we would be back to classical process design rather than evolutionary change management. In this part of the workshop, the team examines the current, *running* processes in the light of the previously identified (aggregated) work item types. Subsequently, after all this theoretical preparation, we finally get our hands dirty with some practical attempts: the team has built the first instruments with which it can test its Kanban system in a simulation.

17.2.1 Finding the Work Steps for Our Work Item Types

How has the team been doing with its schedule in the system design workshop up until now? If there is enough room for maneuver, then, in the name of better quality, we recommend another small-group exercise. The decision as to approach is a question of judgment depending on the time available and the required quality of the results. Do not forget that the team does not have to design a perfect Kanban system but rather one with which it can start working the next day. In our workshops, we always make situational decisions between the following two methods:

1. Each small group describes the current processes for each work item type.
2. The facilitator shares out the work item types among the small groups, and they determine the processes for only the work item types they've been assigned.

The first variation has the advantage that it generates better discussion results and a clearer shared understanding of the processes for the whole group. At this stage, everyone has already considered the same questions, and therefore, it is immediately clear how and where the extant processes are perceived differently by different people. If the participants haven't studied the same content before the presentation, then they will probably more readily agree with what they are presented, without having put too much thought into it. This obviously saves time and we progress through the workshop at a faster rate; however, if there is enough time, honing the shared understanding and the improved end quality that comes with it should be the main focus.

17.2.1.1 The First Board Prototypes Regardless of how you identify the process steps, grab the Post-its and get some pin boards ready for the group work! Later, when fully operational, the Post-its will function as "tickets," that is, as real work items. In the meantime, however, they still represent the individual work steps (e.g., Analysis, Design, and Test) through which work items progress.

With his team, Stefan Bergmüller identifies the work steps for each work item type in the **World Café**. How does this work?

- Firstly, he splits the group into three teams. Flipchart paper lies on each table, along with plenty of Post-its and pens. "I'm going to give each team some of the work item types we identified earlier," explains Mr. Bergmüller. "For each of these work item types, please think about the steps through which they progress and in what order this happens. Then write each step down on a Post-it. We're interested in the actual current process and not a theoretical process we'd prefer."
- After 15 min, Stefan Bergmüller changes the teams; but one team member remains seated and explains to the new arrivals what was discussed and decided upon by the previous team. Thus, the teams tackle the questions in new and different constellations.
- Ultimately, Stefan Bergmüller reshuffles his little World Café groups three times, with the result that the first provisional boards have already evolved. The team members attach the work item types beneath each other on the left (the swim lanes) and running horizontally next to them they order the work steps required to complete the respective work item type. It now becomes clear that Post-its are used for the simple reason that the process design can be more easily rearranged with these removable notes if new results should appear in the course of the group work or later on in the plenum.

USING SCALE SENSIBLY

In this small-group work, one question in particular will occur to many participants: "What counts as a process step and at what point is the degree of detail too high to be of significant benefit?" Well, let's think. In what sort of environment is the Kanban team located? Let us consider a support team and a software development team. And let's also consider that, when operational, a ticket always needs to be pulled into the next process step as soon as a work item has been completed.

A support team usually pulls tickets from the input queue into the done column within 1 day, sometimes even a few hours. Is it therefore sensible to document every process step fastidiously on the board? If the team wants to generate a lot of palaver and reallocate tickets currently being processed, then yes, maybe. But in fact, it's too much of a good thing and the *compilation of process steps* is significantly more efficient, not to mention the nerves it spares!

On the other hand, with a software development team, it can take weeks before a ticket can finally be moved further on. This is more frustrating than stressful because the visualization gives everyone the feeling that nothing's happening. One solution here is to *split the tickets* and therefore reduce their size if possible. This creates more movement on the board and shows that the project is advancing.

How does Stefan Bergmüller's team conclude this stage of the workshop? The groups present the boards they have created and together compare the suggestions. From this consensus evolves the team's first Kanban miniboard. "I'm pretty impressed that you've already integrated a buffer," says Mr. Bergmüller enthusiastically, since he's learned from us that this kind of leap is not typical in the system design workshop. It wouldn't have been a problem for him if it had taken a little longer—ultimately, the whole workshop is a learning process for the team, in which they get to know Kanban properly for the first time or at least in which they get to put into practice for the first time what they acquired in the Kanban basic training. "Sometimes it happens quickly and sometimes it takes a little longer. The main thing is that the team has understood the thinking behind the work," Stefan Bergmüller thinks to himself.

Team Bergmüller takes another, closer look at the board. "Is there anything we need to change?" team leader Bergmüller asks, "Or can we get started with the board tomorrow?" The team members now need to do the following for a structured feedback process:

- **Thumbs up** means, "I'm all for it."
- **Thumbs pointing to the side** means, "I'll do what the group wants."
- **Thumbs down** means, "I'm against it."

Thankfully, the team's not deciding the fate of gladiators, because a couple of the thumbs in the room are pointing downward. "I can see that for you guys, everything's not quite in place. What do we need to change to get your thumbs at least pointing to the side?" Stefan Bergmüller asks. It turns out that both of the team members aren't satisfied with the process for the work item type "Hardware Update" and have a different suggestion for it. The board is altered and Stefan Bergmüller asks the question again: "Can we make a start with this board tomorrow?" Another round of finger voting: not a single thumb is pointing downward; the suggestion has been accepted.

ELECTRONIC OR PHYSICAL BOARDS?

Nowadays, there are of course electronic solutions for Kanban boards. There might come a point when your team asks whether it wouldn't perhaps be more practical to represent your system with a program. Among other things, electronic solutions have the advantage that they deliver measurements automatically, and these can help the team with further improvements. Nevertheless, we are proponents of the board that physically *exists* and most teams with whom we've created Kanban systems agree. Why?

Do you carefully read all the e-mails you receive? Are you one of the disciplined creators of Excel spreadsheets used for all possible purposes? If so, then we congratulate you on your unusual level of commitment. Most people only scan over content, and that's if they open the e-mails at all. It's pretty much the same with electronic Kanban boards: at some stage, certain slackness will set in and you won't follow what's happening quite so carefully. **What electronic boards lack is visibility and presence, and this is precisely what we want to achieve with the visualization of our work.**

In practice, we frequently come across a solution where teams use electronic boards but nevertheless make them visible for everyone in the organization. Screens hang in the team rooms and in the hallways and display the current state of the Kanban board for all to see. Or the team uses a projector in order to have the board permanently on display and not only after clicking around. These are great solutions in and of themselves, but nevertheless, we'd like to give you one more thing to think about before letting you make your decision. *The degree of interaction between team members increases with simple means such as Post-its and cards.* Tickets are taken from the board and discussed in groups. Notes, sketches, or small diagrams are not possible with electronic tools because they force the team members' ideas into a predetermined and limited set of technical functions. And in so doing, technology rather than the people themselves becomes the determining agent of change.

17.2.2 Simulation and Ticket Design

The time has come. The team assembles for the first time in front of its own Kanban board and begins to test whether or not it's practice proof. The team should *not* be split up for the simulation because there are many variations: perhaps each person is responsible for a particular process step or team members work through the whole swim lane for certain work item types. Separation into small groups is thus neither sensible nor possible. At this point, it is important for the team members to be shown how work is distributed among them and to recognize where the handover points within their own group lie.

And that's exactly how support team boss Stefan Bergmüller does it: with his team, he simulates the identified work stages and takes on the swim lanes one after the other (i.e., each work item type). Everyone is concerned about the following question:

"Which work items move most frequently through this swim lane?" Mr. Bergmüller quickly notices that as facilitator for the simulation he will have to bring all his pragmatism to bear. The typical technician-like analytic nature quickly becomes apparent: upon their first practical experience with the board, several team members want to represent all potential cases in order to be safe. Stefan Bergmüller notices relatively quickly how the team's mood goes downhill due to this. "We'll never be able to represent everything that could possibly happen," says one person, sighing loudly. "Neither should you. You don't need to build the most watertight Kanban system of all time," says Mr. Bergmüller, trying once more to encourage them. "This Kanban system isn't there to make us feel depressed—it's meant to help us." He repeats something that's become a kind of mantra by now, "We're not building a system for the 5% of cases that could possibly, perhaps, just maybe happen. We want to design a system for the 95% of cases that we encounter on a daily basis." For every new special case created, Stefan Bergmüller simply asks: "How often does this special case happen?"

Once attention has been firmly turned back to the typical cases, Bergmüller's team takes a real example from the pile of normal work items and writes it on a ticket. The team is not initially interested in why or how this work item gets into the input queue—this will be tackled at a later stage. The team now follows this case through all its work steps as follows:

Preparation > Execution > Documentation

The fact that the steps *Preparation* and *Execution* aren't really separated in the *existing* process is pretty quickly established. "No, that's not true!" claims one employee of many years' standing. "We first need to complete the preparation and then begin with the execution. Look at the process description, it says so there." Stefan Bergmüller obviously can't just ignore this objection. "Well give me an example with one of your work items, and we can run through the process with it," he replies. Once the example has been chosen and the simulation run, it turns out that for tickets that are already in *Execution* stage, there are always elements of the preparation phase that must still be carried out. In practice, therefore, the ticket must either be blocked—since preparation work still needs to be done retroactively—or it simply gets returned to the *Preparation* step—what's more, several times over (which is definitely *not* our favored variation). With the help of this visualization, it becomes clearer to the slightly critical employee that the only sensible option is to merge the two steps into one step called *Preparation and Execution*.

It's also very likely that your own system design workshop will need a bit of reconstruction in the course of the first simulation. Some work steps are removed, and others integrated. It is often only during the simulation that it becomes clear that the work item types can be grouped together even further. *Keep repeating this reworking process until the team is standing in front of a board with which they can really begin working.*

In order to gradually improve the system and align it even better with reality, the facilitator must steer the group thinking a little. In our workshops, for the simulation of each swim lane, we always ask the team which blockers might typically appear.

With this question, we guide the thought processes to the issue of whether further queues or buffers need to be incorporated. This is often the case if work items are handed over to external suppliers midway through the process or if they need to pass through an internal review process out of the team's control. Even if we do seem to be focusing a lot on design, the (initial) final Kanban board hasn't yet been born.

When we ask about the example job we should use for the simulation, another question often immediately follows: "What does it actually say on the ticket when it arrives in the input queue?" Time and again, the topic of ticket design appears during the definition of the work item types, when people are unsure of things. But it is more often the case that things reach a head during the simulation because it's now necessary to commit something to writing. The creativity of the team isn't given any limits here: anything is allowed as long as it is good for collaboration and delivers information that the team can later use for its measurements. In Chapter 3, we saw how a ticket can be developed.

SUMMARY AGENDA POINT 2: IDENTIFYING THE PROCESSES

Marking out the framework

The facilitator reminds the team that it is the *existing* process that is to be represented and not a desired or prescribed one.

The process

1. Description of the process steps for the individual work item types in small groups:
   a. Possible modeling method: "World Café," where the groups are mixed and remixed, generating a more comprehensive perspective of the issue
   b. Result: first board designs
2. Presentation of the designs in the plenum and their aggregation into a shared, provisional Kanban board
3. Simulation of some typical cases from day-to-day work: compilation, splitting up, or addition of process steps

The board is altered and tweaked as long as is necessary for it to represent reality and so that work with it can be started the very next day.

17.3 DETERMINING THE WiP LIMITS

Up until now, the first part of the workshop has strongly focused on the level of visualization. This is important groundwork because it is only through this that a team becomes more aware of itself and actually *sees* the work items it processes and how it proceeds with them in the existing process. Only when the team fully understands the current situation can it estimate how it can use the other Kanban instruments, those that will highlight the weaknesses in the process and thus help to complete work items more quickly and reliably. The most important instrument in this context is the WiP limit. Let us think back once more to our understanding of WiP limits and the effects they have.

The goal of a Kanban system is to establish a continuous workflow. This won't work if uncontrolled work items are constantly being poured into the system. We therefore limit the number of work items that may be passing through a certain process step at the same time. Foremost is the economic concern that a completed work item is more beneficial than ten work items that have been begun but not completed. WiP limits have the following effects:

- Switching between work items is avoided.
- Lead times are reduced.
- A higher level of quality is achieved, thanks to better concentration and shorter feedback times.
- Customers can better rely on delivery estimates.
- The team is disrupted far less while working.
- Problems and bottlenecks are identified via "queues" in process steps.

Throughout the process, the facilitator and the team should bear in mind the results of the review workshop and interview compilation. In all likelihood, several of these points will crop up in the list of desired improvements, if not verbatim, close enough. Let's address these wishes now by providing the team with the Kanban board they have just created!

17.3.1 Step 1: Finding the Right Capacities

The facilitator finds an entry point to the topic via the swim lanes that represent our work item types. The simple question (with no simple answer) is: "Roughly what percentage of your work time is spent on the individual work item types?" In the workshops that we carry out, we put this question to the plenum and leave it up to the team to choose how they would like to organize themselves to provide an answer. There are two possible approaches to this:

1. Relevant records have been kept for time management purposes or for other reasons. Very practical! Now, we can make use of these records in order to answer the question of percentage distribution precisely.
2. There are no records. Answering the question will take a little longer, but this isn't really a problem. A team is usually capable of making a qualified estimation based on experience. Moreover, by now, it's become instilled in everyone that we're not trying to build a nuclear power station but rather a pragmatic Kanban system that is forgiving of errors because it can be adapted so quickly.

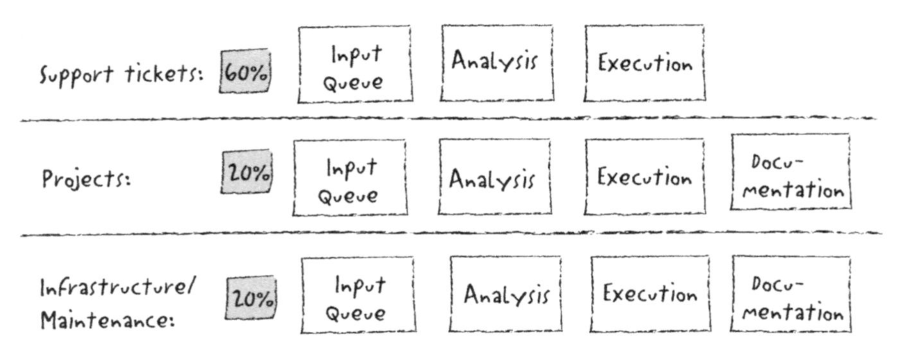

FIGURE 17.1 Distribution of percentage proportions for the individual work item types.

If there are no records, the team members write their estimates on a flipchart and calculate a simple average of all their answers to generate a (rounded) percentage for each work item type. Figure 17.1 shows such a chart.

In the system design workshop, most teams are convinced that there is a clearly defined phase called "analysis" in their process. However, in daily operations, it often turns out to be the case that *Analysis* and *Execution* actually overlap quite a bit and are fundamentally not separable into discrete steps. We make suggestions in the workshops, but we accept that the team may see things differently. However, the teams normally come around to our opinion within a week or so.

This stage is often full of big surprises. One IT infrastructure team we worked with wanted to invest more time in the improvement of the internal IT infrastructure in the future. This was to constitute at least 50% of the work done for all work items. An examination of the current status made it clear that achieving this was still a long way off. The amount of time currently spent on internal infrastructure issues was about 20%, while 30% went into project work and the remaining 50% went into support requests. Such insight might come as a shock at first, but it then activates the creativity of the team: "What do we need to do to this distribution so that we can reach our goal? What might an optimal distribution look like? What impact would turning a particular screw have? How do the individual work item types influence each other in our field?" In our current example, the intense concentration on improvements to the internal infrastructure would have resulted in the number of requests via the support line decreasing. The team decided to opt for a gradual alignment of capacities.

We can recommend such an approach not only for this example but also generally, too. If the capacities are instantly radically redistributed, this would immediately overwhelm not just the team but also the other stakeholders, who would find themselves confronted with a totally new situation overnight. Evolutionary change isn't just a matter for the Kanban team but also for all stakeholders who have a connection with the team.

At the end of step 1, the team agrees on a capacity distribution that is currently sensible, feasible, and compatible. It then records this conclusion on its first board.

17.3.2 Step 2: Translation into WiP Limits

Initially, the team took a bird's-eye view of its system: "What does our system achieve as a whole?" A system can only work optimally when its intake ability is considered and accordingly limited. We therefore now need to switch to the individual team members' perspectives and, using these as a basis, determine the WiP limits. In Section 4.2, we said that there is no predetermined "right" WiP limit:

- Firstly, all we need is a starting point to begin with the practical work. It is only in actual operation that the team will realize whether the limits for the start phase were set too optimistically or pessimistically. Naturally, the WiP limits will change over the medium to the long term because the team will perform major development steps.
- Secondly, the team needs to gain experience through observation and adaptation. The goal of WiP limits is the establishment of a constant workflow by reducing work items' "sitting time" in the "done" column. This reduces the lead times. However, the best way to show how this can be done is through experience and the help of an intimate Kanban know-how gleaned over time.

An easily comprehensible starting point therefore needs to be established. To achieve this, the facilitator of the system design workshop asks the following question: "How many work items can each of you *realistically* process simultaneously?" The number of employees is used as a measure. The team members submit their estimates based on experience and self-knowledge. We now have the ingredients we need to calculate the WiP limits. We already showed how this works or could work in Section 4.2:

$$\text{Number of people in the team} \times \text{estimated number of simultaneous work items possible per person} = \text{WiP limit for the whole board} \quad (17.1)$$

For the sake of simplicity, let's assume that we are working with a Kanban team of five employees who are each capable of processing two work items at the same time. We thus have a WiP limit of 10. This limit applies to the whole board and all swim lanes (see Section 17.2.2).

Using the example in Figure 17.1, let us look at how the various proportions accorded to the work item types hitherto expressed as percentages are converted into WiP limits. Incidentally, we consider this part of the workshop to be very short and painless—we simply facilitate the reformulation. Ultimately, we're not dealing with the most cutting-edge mathematical techniques but rather a simple calculation that every team member is capable of. The WiP limit of 10 is our 100%: 60% of 10 is 6; 20% is 2. Figure 17.2 shows the board updated with the WiP limits.

You have probably already realized we're not quite finished with the WiP limits. Naturally, the limits also need to be distributed across the individual work steps—this is clearly a task the team must tackle. The team needs to decide, for example, how it distributes the six support WiPs across the work steps *Analysis* and *Execution* in a way suitable for the demands of everyday practice. If you cast your mind back to

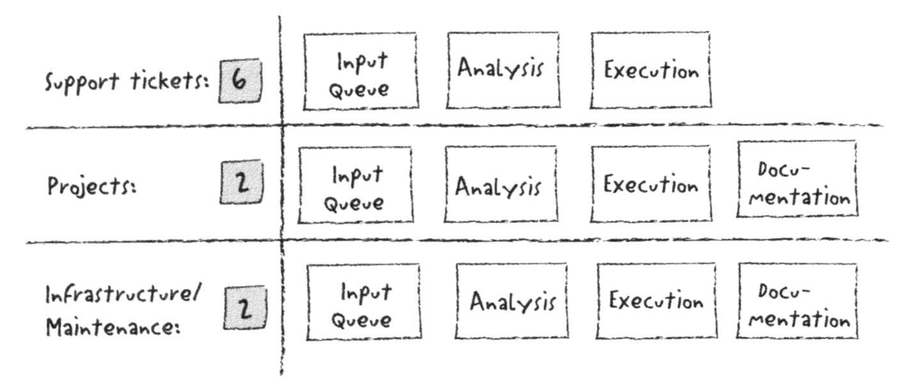

FIGURE 17.2 Conversion of work item-type percentages into WiP limits.

Section 17.2.1, the team was still working with a board on which the individual work steps were visualized with Post-its. The team simply sticks the WiP distribution suggestions for the various individual WiP limits onto the relevant work step sticker. At the end, the distributed WiP limits per work step are added up and the total is written on a new Post-it and stuck above the respective work step column. It's worth checking again that you haven't allocated more WiPs than you have actually available. If everything fits, the team now has developed a Kanban board that is almost ready to be deployed!

Let us run through the whole process one more time, this time using the team Head of Development Susanne Schweizer leads as an example. Before finding out its capacities and setting the WiP limits, her team was getting bogged down in special and unlikely cases in the simulation. However, as facilitator, Ms. Schweizer had succeeded in directing the concentration of her team members. The translation of the capacities into WiP limits therefore proceeded quickly and without a hitch. This is probably also because at this point she made the material much simpler. She knows that the team should actually orientate itself around the work items' waiting times in order to set the WiP limits. However, she also knows that her people have already done a great job so far and that the concept of flow optimization would be overwhelming at this stage. It takes some time for the team to come to terms with Kanban and change their perspectives bit by bit. At this point, determining the WiP limits using the number of employees is the more enlightening method. What process steps are available to Ms. Schweizer's team (Fig. 17.3)?

Ms. Schweizer's team has already incorporated a buffer called "ready for testing" on their board, in order to absorb variation. They did this because while reflecting on the process steps, the team members established that the steps *Design and Development* and *Test Development* didn't typically take the same amount of time. Experience tells them that this buffer is thus necessary in order to avoid bringing the workflow to a halt. They can't yet say whether or not they're ultimately really going to need the buffer, but they start with it in place. If the buffer is unnecessary, they can remove it at a later date.

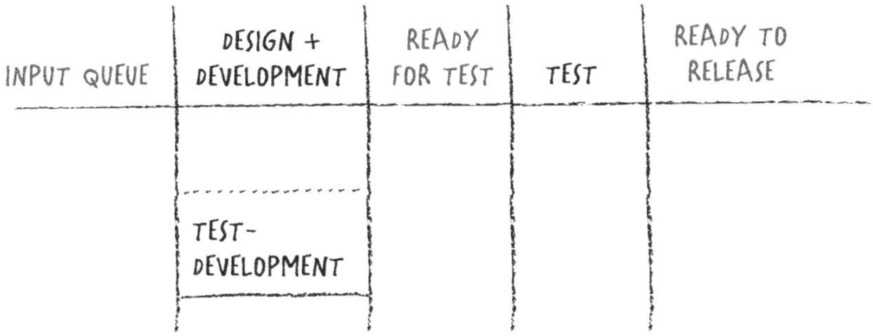

FIGURE 17.3 Kanban board for Susanne Schweizer's development team.

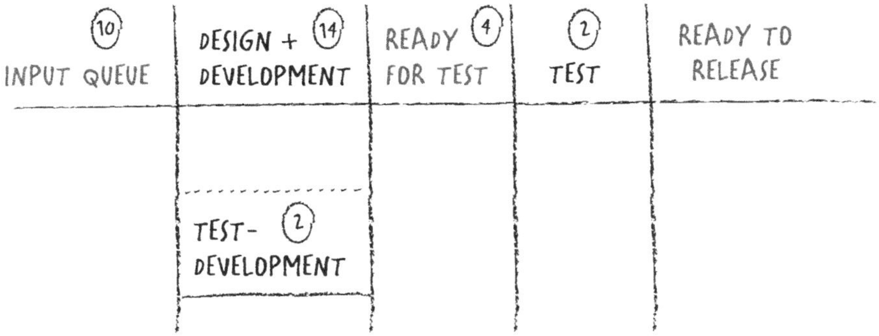

FIGURE 17.4 Team Schweizer's Kanban board with WiP limits.

The team consists of nine people in total: seven developers and two testers. "Please think about your work situation," Ms. Schweizer begins. "What do you believe? How many work items can each one of you reasonably process?" The team agrees on a figure relatively quickly: they decide that any more than two simultaneous work items per person would be counterproductive. Figure 17.4 shows how the WiP limits are distributed across the development team's Kanban board.

Numerically, the developers have an overall WiP limit of 18 *for the work steps*:

- The step *Design and Development* is restricted to a WiP of 14 (seven developers; two simultaneous work items each).
- One WiP limit of 2 for *Test Development* and *Test*, respectively (two testers, two simultaneous work items each).

After a brief discussion whether or not the buffer needs a WiP limit, the team gives it an off-the-cuff WiP limit of four, since they still need to gain the relevant experience. For the time being, it's clear to the team that the buffer also needs to be limited so that work items don't simply pile up there. The WiP limit of the buffer does indeed raise the WiP limit for the whole board, but the WiP limits for the individuals remain unaffected. This is because the buffer is actually only a "car park" for work items that are currently not being processed.

The input queue receives a WiP limit of 10. This WiP limit regulates the inflow of work items into the Kanban system. How does Ms. Schweizer's team reach this WiP limit for the input queue? Before the Kanban initiative, the wishes of those responsible for the products were delivered to Ms. Schweizer's team in 2-week intervals. "But we all know that it never stays like that," a team member observes, with a slightly strained smile. "We always have these requests, people asking us if we could perhaps just do this or that work item really quickly." Thus, it is decided that the queue replenishment meeting should be held on a weekly basis in order to capture and better channel the wishes that evolve outside the meetings. With more precise observation, the team manages six user stories per week. Sometimes, it's as many as nine and sometimes as little as four. "Do we want to shape the input queue in such a way that there is always enough feed available?" Susanne Schweizer asks. The team is in favor and the WiP limit for the input queue is therefore initially set at 10 to err on the side of caution. This limit is accompanied with a resolution to observe the next queue replenishment meeting carefully and, if necessary, reset the limit to a lower level if 10 would turn out to be too high.

17.3.2.1 Determining the Size of the Input Queue We know from Kanban theory that the size of the input queue depends upon the throughput of tickets in the system and the frequency of the queue replenishment meetings. The question is whether one of these variables is already known. In most cases, the frequency of the queue replenishment meetings has not yet been decided in the workshop, but it very often is the case that they take place on a weekly basis. The team will tackle this in greater detail at a later point in the workshop. Since these meetings are still to be introduced, therefore completely new, the cadence can easily be determined right now. But what can we say about the approximate number of tickets a team can work through per week (throughput)? The point is that a team doesn't yet know this; it will only become clear during constant operation—therefore, answers will have a high degree of variability. The team might manage five tickets but it might also manage 20. It depends on the tickets.

Just to remind you: right now, we're creating a Kanban system with which we can get started. We therefore ask our teams to *initially* err on the side of caution in terms of this estimation. "We believe that we can cope with x tickets per week, maximum." We set the input queue to this level and, during operation, observe how quickly it is emptied. As soon as the work items run low, we invite the stakeholders to a spontaneous, advanced queue replenishment meeting because we know that the size of the input queue needs to be raised. Alternatively, if the input queue is still full, the team needs to put the queue replenishment meeting off until a later date and reduce the number of work items waiting.

In practical implementation, the initial problem is normally that far too many tickets get piled onto the board because the team also needs to finish leftover work items over from the pre-Kanban time. The primary task is therefore to empty the board until all are working within the WiP limits and go without queue replenishment meetings for a while. This period in fact generates a very good idea of a suitable size for the input queue.

17.3.2.2 When Work Items Depart From the System We're sure you won't get the impression that setting WiP limits in real practice is as smooth and trouble-free as we've just seen. Of course, it's not. This part of the workshop is usually the one in which the team members become most aware of how dependent they are on other units in the organization and of the limitations of their own workflow. This question often initially appears during the identification of the work steps, but sometimes, it is only when things are operational that it becomes clear: "What do we do if a job for some reason leaves or needs to leave our Kanban system and therefore our workflow?"

The question now takes on a new dimension: "How do we go about setting WiP limits in cases of external blockers?" In the extreme, the work items disappear in a black hole that also sucks in the WiPs and appear in the Kanban system only at a point when it takes great effort for the team to affect them if at all. Generally speaking, we regulate this temporary handover of work items to other areas of responsibility through the introduction of queues. We thus emphasize the effect of these temporary handovers: namely, delays through waiting time.

At the stage in a process (or on a board) when work items leave the system at undetermined times, we set up a queue that serves as a ticket car park. This queue receives its own WiP limit.

To see this, let us take another look at the case of Susanne Schweizer's team. In the first retrospective during constant operation, the team establishes that it overlooked two dependencies within the organization in the system design workshop. Before the software can go into the test stage, the team often has to wait for an infrastructure team to prepare the required server environment. Then there's also a second development team that needs to integrate parts of its work with the work done by Ms. Schweizer's team. In retrospect, team Schweizer expands their Kanban board to include the "waiting for external events" queue (Fig. 17.5).

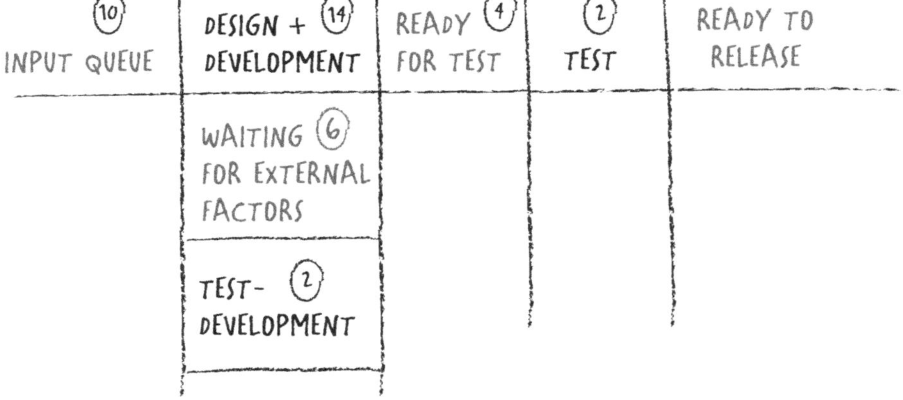

FIGURE 17.5 Handling external blockers with queues.

To begin with, the team limits this queue to six tickets. *Theoretically*, there would be a total WiP limit of 38 for the whole board. *Practically speaking*, however, this doesn't mean that the team suddenly takes on more work than previously. The distribution of the WiPs across the work steps remains the same as before, since there is no increase in the capacity the team has available.

A queue is not included in the distribution of capacity, be it the input queue or a "car park" queue.

The team achieves two things with the WiP limit for the queue:

1. When the number of tickets in the queue gradually approaches the limit, the team will immediately see that it needs to request further work items.
2. Due to the WiP-limited queue, every colleague or stakeholder that walks past the board will see that the team is in waiting mode. This could well become the impetus for further structural improvements. Incidentally, just before this book went to print, we heard from Ms. Schweizer that her team had been combined with a second team and that the process now runs significantly quicker and smoother than before.

In system design workshops, external blockers are usually the cause of intensive discussion on how the WiP limit for this kind of queue is supposed to be applied. Ultimately, a team doesn't know when it will get work items back or when management will make a decision based on which they can pick up a work item again. Can you even introduce a limit for this kind of queue? Are you allowed to? Our answer is that with this kind of queue, a WiP limit *must* be introduced. The level of this limit is just as much a learning process as in the case of normal WiP limits. Leaving the queue unlimited would mean handing control of the lead times to external agencies. And then, when the work items are returned, the team would be "allowed" to start working again. But with a WiP limit, the team is also forced to follow up in good time. WiP limits for a queue are thus an indirect means of communication with stakeholders. With time, those outside the team will begin to appreciate that mutual dependency very much means being responsible for the other party.

17.3.3 Distribution of the WiP Limits for a Support or Test Team

If, for example, a workshop is being carried out with a support or test team, there are three specific issues that will most likely emerge:

1. Firstly, with these teams, it is very often the case that one employee will work alone on a ticket, right through all the process steps.
2. Secondly, many tickets will have very small lead times often lasting around an hour.
3. Thirdly, these teams will have many different work item types to complete.

Six work item types were identified by Stefan Bergmüller's team, which is responsible for second-level support. In 90% of the cases, a ticket was processed by only one of the seven team members. In a team of seven with a maximum WiP of two tickets per person and the additional fact that one person carries out all steps per work item-type ticket, it doesn't make much sense to distribute the WiPs across the individual process steps. Thus, Stefan Bergmüller's team restricts the WiP limits to the work item types.

Stefan Bergmüller's team processes 20 tickets through the Kanban system every day. He knows what can happen in this situation and therefore simulates the work process for some tickets with his people. Pretty quickly, they establish that they can undermine the distribution of the capacities across the work item types. In the swim lanes with high WiP limits, work items are suddenly parked (or simply stay put) while the work items in the swim lanes with lower limits are advanced quickly across the board. The team is here falling victim to the temptation of self-deceit. Sure, its WiP limit is always at full capacity and completed work items are still leaving the system. But it's always the same kind of work item types that are being completed quickly. In order to nip this temptation in the bud, Team Bergmüller decides to limit the capacity distribution process to a weekly basis. It also asks itself the following question: "How many tickets can we complete per week per work item type?"

Twenty tickets per day means around 100 tickets per week. Based on this, the capacities are distributed anew and the team also establishes how many tickets per work item type are allowed to be on the board from Monday to Friday, *including the "Done" column.* As a rule, the team stipulates that the tickets must remain in the "Done" column until just before the first daily stand-up of the following week, that is, Monday, when the "Done" column is cleared out and can therefore be filled again. In order to get a feeling for whether this solution might work, the team simulates the flow once more based on the new policies. They discover that there is no longer any danger of deceiving themselves.

SUMMARY AGENDA POINT 3: DETERMINING THE WIP LIMITS

Marking out the framework

The results from the interview compilation in the review workshop are used to determine what we want to achieve with the WiP limits.

The process

1. Finding the capacities: What percentage of the working hours is spent on the respective individual work item types? Possible methods of enquiry:
   a. Records from the past
   b. Qualified estimate based on experience

2. Translation into WiP limits: How many work items can each team member realistically process simultaneously?
   a. Using the percentage distribution from Point 1, the result is converted into WiP limits for the individual work item types.
   b. Distribution of these WiPs across the work steps. The input queue, any other queues, and all buffers are excluded from the distribution. They receive their own WiP limits, generated based on experience and advance estimates.

17.4 DETERMINING THE CLASSES OF SERVICE

The last things we need in order to efficiently control the lead times of our Kanban system are the classes of service. A brief flashback to Chapter 5: *Classes of service are combined with a cost of delay. This cost of delay elucidates the consequences of completing a work item too late or not at all.* By separating them into classes of service, work items are differentiated from each other according to urgency. Thus, a team can better control their workflow and assign the work items the necessary capacities.

In Chapter 5, we introduced the most common classes of service. We will repeat that, through practice, we have come to the conclusion that the number of classes of service must be manageable; otherwise, the team members and stakeholders would have to maintain a great tangled mess of policies. Six or seven classes of service are probably already a bit much. Yes, having a lot of classes of service means you can measure the workflow right down to the least detail, at the cost of control and ease of understanding. Initially, therefore, the goal is to focus on a few classes of service: four would normally suffice, with the number later increasing to six at most, if necessary. Initially, the team should represent the standard cases in order to gradually develop a constantly improving understanding not only of its own process but also of the processes of the organization as a whole. It is the classes of service that will increase awareness of the role one's own work plays in the success or failure of the organization. It will become clearer that managerial decisions are not just arbitrary acts but decisions that have their causes in strategic reflection. However, when you want to leave the role of someone who simply receives orders to become someone who actively changes things—which is what the members of a Kanban team do—you must gradually become acquainted with these complex relationships. Of course, teams that begin with Kanban do not all have the same level of "maturity," but nevertheless, with the definition of the classes of service, we ask that you begin with what is simplest and happens most often (as is indeed the case with all other elements). Most important is that the team members really and truly understand the concept.

This is the reason why we work with such simple tools as Post-its, which fortunately are available in a range of different colors. *Don't forget: the team needs to decide which classes of service have which colors and whether a class of service under certain conditions should perhaps even receive its own swim lane.* The questions as to the classes of service you have, the policies that govern them, and the colors they're assigned will also become a subsequent topic of communication with the other stakeholders. Ultimately, as contractors, they need to understand how our system functions so that they can orientate their own actions accordingly.

Let us see which classes of service Susanne Schweizer's development team has chosen. A very productive discussion has started here, where many financial thoughts and ideas have been brought into the mix. Ms. Schweizer observes how her employees genuinely begin to think of how their own work impacts the market opportunities of the organization. This was precisely what the head of development wanted to achieve with her Kanban initiative. Her problem was that the team was constantly tied up in trivialities and product management had already lost its patience. Everyone immediately agrees that there needs to be an expedited class of service. Firstly, this is because major errors repeatedly appear in the productive system and have to be eliminated quickly. It also becomes clear that the software quality definitely has to improve, but this can't just happen from 1 day to the next. Secondly, product managers frequently need speedy support from the development team because winning a new contract depends on it.

This constant preoccupation with quality issues prevents the team from developing new features. It therefore defines a second class of service called "necessary to stay in business." This class is for key features already offered by the competition and therefore requiring swift delivery in more or less the same form.

However, the team now wants to create the situation whereby it doesn't just keep pace with the competition but rather overtakes them. Therefore, the class of service "features for winning new market share" is created for the features with which they intend to win over new users.

17.4.1 Creating Clarity between Work Item Types and Classes of Service

In order for classes of service to work properly, a Kanban team needs to accompany every class of service with a set of policies. However, it is very often the case that while working on the classes of service, the work item types already identified, that is, the swim lanes, are fundamentally altered once again. During this phase, it could also be the case that classes of service become their own work item types and that original work item types become classes of service. When considered carefully, classes of service aren't something we're "inventing," pulling out of thin air. Classes of service already exist in the team's daily work—it's just that the team isn't aware of it. *Before* Kanban, the team also had different requests and handled them differently. However, in this case, deadlines and pressure applied by stakeholders are probably the determining factors, obscuring the fact that there are other more organized ways of achieving the goals.

These classes of service, already implicitly present, are hidden in the individual work item types. In our workshops, we always establish whether everyone has understood the concept of cost of delay right at the start. Once all uncertainties have been put to rest, we invite everyone to reflect once more on the work item types on the board. The actual request is this: "For every work item type, think of the impact

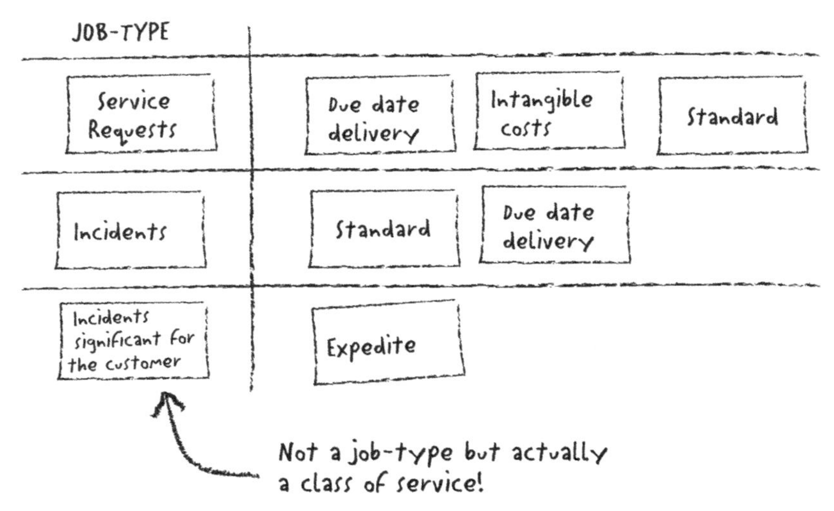

FIGURE 17.6 Work item types versus classes of service.

that work items falling under this type can have on the business. Try and represent this impact as a cost-of-delay graph or describe it in your own words." By saying this, we are not restricting people to a graph but giving them the option to use more descriptive methods. In doing so, there is a better chance that relevant risk information is mentioned, since this cannot easily be squeezed into a graph's somewhat narrow corset.

The various diagrams and descriptions for each swim lane are now collected. Figure 17.6 illustrates this process using the Kanban board of IT Head Josef Drechsler's team. The preceding steps in his team's system design workshop had suggested a total of four classes of service in the three work item types: "incidents," "customer-relevant incidents," and "service requests." The team uses conventional names for these four classes of service.

As the team takes a closer look at the board, it notices that all the work items in the swim lane "customer-relevant incidents" are expedited work items, which isn't the case at all with the other work item types. This leads Josef Drechsler to ask the following question: "Are 'customer-relevant incidents' really their own work item type or could we perhaps categorize them as a class of service in one of the other work item types?" If the verdict were, "No, it's definitely its own work item type!" then the swim lane to pull their expedited work items through the system would be retained.

But that's not what happens. Let us take a closer look at Team Drechsler's work item types:

- *Customer-relevant incidents*: these requests have the highest immediate impact; if they are not responded to immediately, the organization is threatened by major financial loss.

- *Incidents*: high but not immediate impact.
- *Service requests*: daily requests from colleagues that, for the sake of customer satisfaction, must be taken care of quickly but not immediately.

After a period of reflection, Team Drechsler decides to amalgamate the customer-relevant incidents and the normal incidents into one work item type and leave service requests as a separate work item type. The reason for this is that when being processed incidents pass through the same work steps, while service requests have their own process. There is now therefore *one work item type* called "incidents," and it has *two classes of service* (customer-relevant incidents and normal incidents). These classes of service are dealt with in the incident swim lane and nowhere else. Had service requests also had the same workflow, then they would also have been amalgamated into the work item type *incidents*.

In summary, then, we can say the following:

- A class of service can also be a work item type. This can be seen in the swim lanes, because these classes of service need to follow their own processes that cannot be combined with other work item types.
- Work item types that were originally separate can be combined into a single work item type if the cost-of-delay graphs for them are similar. In doing so, the originally separate work item types become classes of service.

17.4.2 Defining the Policies

Deciding on and amalgamating classes of service and work item types is probably one of the most intensive steps in the system design workshop. However, once it's done, we can then get started on identifying the team-specific policies for the classes of service. For this, we need to consider two issues:

1. What does the **handling** of the individual classes of service look like? How does the team need to react when it receives a job of a particular class of service? When a decision needs to be taken as to which work item is to be dealt with next, which one is prioritized and for what reasons? In other words, what does the pull sequence look like?
2. **Capacities** need to be reserved and/or limited for the individual classes of service.

In order to discuss handling, let us look at Josef Drechsler's team's solution. As we know, this team defined customer-relevant incidents as a class of service in their work item type "incidents." Together, the team members agreed upon the following handling for customer-relevant incidents:

Whenever a customer-relevant incident that can emerge independently of the queue replenishment meeting appears in our system, the team members from the affected areas put their current work on hold, gather around the board, and think about how they can resolve this incident as effectively and quickly as possible.

With this statement, the team has intentionally allowed itself the possibility that a customer-relevant incident needn't always take up the capacity of all the team members. Half of the team works on the UNIX platform, and the other on Windows. Thus, it is made clear in this statement that only the half of the team whose platform is affected by the problem should concern themselves with the emergency. Within the policy for incident handling, therefore, there is the further situation-based, implicit policy as to which half of the team is to drop their other work items and which half can return to their work after the first few moments of emergency.

Of course, policies can be formulated in a much more restrictive manner if it makes at all sense. Independently of our example, it's probably the case that in almost every software development team, "Features" would appear as a work item type. There would then be several classes of service within this work item type, ideally one for the category "intangible" too. One possible policy might state that the person currently working on a work item of the category "intangible" *must* stop working on it straight away if an error in the production system lands in the input queue.

17.4.3 Establishing the Capacities of the Classes of Service

Regardless of the level of restriction or openness with which the policies concerning handling are formulated, we *must* allocate capacities to the individual classes of service. In Chapter 5, we showed the effects of the most prevalent classes of service on the lead times of work items within a Kanban system. If work items of certain classes of service, particularly those of the "expedited" class, are freely allowed into the system, there is a danger that sooner or later the system will buckle and all its capacities will be focused on this class of service alone. This is what we want to avoid; otherwise, this class of service will ultimately suffer too. We therefore allocate capacities to the classes of service: we establish the maximum number of work items that may be in a class of service in the Kanban system at any given time.

In order to attain a capacity distribution, the team needs to ask two questions:

1. What proportion of our current work is taken up by the individual classes of service?
2. What proportion of our current work should *ideally* be taken up by the individual classes of service?

In the course of the review workshop, several problems the team would like to address with the help of Kanban were identified. Should the team pose the above two questions, individual problems would have another chance to come to light. It is usually the case that "expedited" tickets and the "intangible" classes of service are the most significant candidates here. This is also the case for IT Head Herbert Krakauer's team. They suffer from chronic overburdening due to constant reprioritization and repeated disruptions. "The current proportion of expedited tickets is simply far too high," complains one team member. "At this point we feel like we're simply trying to put out a fire." Herbert Krakauer uses the system design workshop to research the possible causes. "All these expedited tickets constantly hold us back from all other work items," he says, offering an explanation. "The customers have told us often enough that they're not entirely satisfied with our delivery estimates."

There are two possible reasons for the overburdening with expedited tickets:

1. In the past, stakeholders experienced a situation where "the only way of moving anything on" was by making their requests appear more important than they actually were. There was a lack of trust because the team couldn't reliably define the processing period, due to the previous lack of restriction in the system. The alternative is that stakeholders prioritize their work items higher than necessary out of principle, because they are aware of the fact that their work item is in competition with others.
2. The actual problem has its roots in a completely different area, out of the team's area of influence, even though it experiences the effects. In such a situation, the team or a representative of the team must attempt to discuss the situation with the relevant people in the organization in order to alert them to the problem and ideally resolve it.

Case 2 is a matter for deeper causal analysis. In this instance, there is a good chance that the team must continue working in "firefighter" mode until the actual cause of the problem has been identified and the problem resolved. Capacities must then be reserved for this continued firefighting. Let us suppose that a Kanban system is given a WiP limit of 10. A brief look back at Chapter 5 shows that expedited tickets can trump the WiP limit. Let us further suppose that the proportion of expedited tickets in the team's workflow is far too high, lying as it does at 50%. If the WiP limit were to stay at 10, then the team would constantly be burdened way above its performance capability and sooner or later would have to completely give up working on the regular work items. A more productive option would be to halve the WiP limit for the whole system, that is, reduce it to five, until the problem has been resolved, and then devote the other half to the firefighting until the situation has improved.

However, Herbert Krakauer's team's situation falls under the umbrella of **Case 1**. In the past, delivery estimates weren't always honored and, since every stakeholder has a legitimate interest in completed work items, these days they all go for the supposedly certain route of higher prioritization. "I think we need to think about the policies for our classes of service," Mr. Krakauer suggests to his team. "In the queue replenishment meeting, we will talk openly with the stakeholders about the problems and together we will establish appropriate policies." After the first QRM, this policy is as follows: an expedited ticket doesn't need to go through the queue replenishment meeting, but at least two further stakeholders must approve, after brief discussion, that this ticket may be processed before all others. The team of course wants to offer the stakeholders a good service—for the organization's sake, when something's *really* urgent, it wants to resolve it straight away. But this only works if everyone is on the same page in terms of what is really urgent and what can in fact wait a little while.

TIP FOR ESTABLISHING THE CAPACITY POLICIES

- In the case of *lower classes of service* (i.e., those involved with long-term development), a team must sometimes deceive itself. It is very tempting to put these work items off until later because they don't quite have the sense of urgency they might otherwise have. But in actual fact, they are essential for the organization. We need to formulate **Must Criteria** for these classes of service so that some of these work items are always on the board.
- In contrast, *high classes of service* (particularly "expedited") have a very strong effect on the workflow and can bring the work being done on other not-so-urgent work items to a halt for a period of time. We need to formulate **maximum criteria** for these classes of service, for example, "Only one expedited ticket may be processed at any given time. The WiP limit for this class of service is therefore one."

The determination of the class of service capacities can in fact be carried out very quickly, depending on the current situation of the team. However, it could also lead to lengthy discussion with regard to the current problems. Allow this discussion to take place. It is very often the case that causes can be identified and problems resolved during this discussion.

If the team has contributed its suggestions for the classes of service and the distribution of the capacities across them, then it's time for another simulation. At this point, the Kanban system is to all extents and purposes in place and, with the classes of service that have been determined, the team can very easily comprehend what working with Kanban means.

SUMMARY AGENDA POINT 4: DETERMINING THE CLASSES OF SERVICE

Marking out the framework

Concentrate on just a few classes of service—it is more important that the team understands their economic impact.

Program points

1. Differentiating the work items. How can work items be categorized on the basis of cost-of-delay functions? Allocation of classes of service to the work item types and the swim lanes in which they can appear.
2. Clarifying the difference between work item types and classes of service. Can work item types be amalgamated? Should certain classes of service become their own work item types?

3. Defining the policies. How should the team react when it receives a work item of a certain class of service?
4. Establishing the capacities. What proportion of the work currently consists of the individual work items and what proportion *should* consist of them?Definition of minimum and maximum criteria.

17.5 DEFINING THE MEASUREMENTS

In Chapter 7, we had already emphasized that you should not treat the measurements of the Kanban system's performance scientifically. You can of course measure things down to the last detail, but you should ask yourself one question: "Are we measuring because it is producing new insights, or are we measuring simply because we can?" It is due to the dubitable relevance of excessive measurement that we advise the following:

> *In this part of the workshop, restrict yourself to whatever is significant and useful: at this point, you are only lighting the fuse.*

What is more important is that the participants have properly understood and can use all the Kanban instruments. Measurements are a small part of a Kanban system. Our experience shows us that in the system design workshop, it is necessary to once again stress the goals that can be addressed using the measurements. During actual operation, measurements suitable for the team will develop of their own accord.

But we do have good news for you. The people with whom you will hold the workshop are not going into it blindly. They either have already been involved with Kanban before, either during a training course or working with it, or are getting to know Kanban "on the fly." At any rate, the participants constantly develop a better feeling for Kanban: they begin to understand it properly and consider several aspects simultaneously in every step of the workshop. Therefore, the issue of measurement of reliability, for example, can already appear during the identification of work item types. While setting the WiP limits, the team will concern itself more seriously with lead times and perhaps even with the issue of how they can accompany changes with specific measures.

It is therefore the workshop facilitator's job to indicate to participants that the subject of measurements is being touched upon at these points. Reserve a flipchart for measurements throughout the duration of the workshop and stick to the points that are mentioned regarding this topic. It is likely that you will have already collected a wealth of potential metrics by the time you reach the agenda point "measurements."

When it's time to select the appropriate measurements, give the team the following advice:

- **Measure whatever helps us reach our goals**: A team uses a Kanban system to pursue certain goals defined in the compilation of the review and interview topics. We are therefore not measuring just for the sake of measuring, but rather in order to learn whether or not we have approached these goals. Measurements help when they show where optimizations can still be carried out.Let us return to Stefan Bergmüller and his support team for a positive example. Naturally, honoring the service-level agreements is a particularly important point for a support team, and this is precisely what was right at the top of the stakeholders' wish list. Therefore, the team members decide to begin by concentrating on one single measurement: that of the lead times of the individual work item types. Everyone agrees that every time the handmade measurements are transferred into an Excel spreadsheet, a different team member should be responsible. The Excel spreadsheet automatically calculates the throughput and the histogram can be generated at the press of a button. The team can then define service-level agreements based on this information.

- **Quality of statement instead of quantity**: When just starting out with a Kanban system, a team shouldn't be overly concerned with having lots of measurements but rather ones that really say something about how the changes have impacted the workflow.
- **Watch out for interdependency**: Changes need time. It is for this reason that they only become visible after a certain amount of time. But this also means that you shouldn't resolve to undertake too many changes all at once. It will become no longer possible to tell from the measurements which change is responsible for which result. Moreover, having too many simultaneous changes can lead to them hindering each other's effect, meaning that the team fails to achieve the desired effect.

RESPONSIBILITIES

So far, all Kanban system elements have evolved from a combined effort by the team. Actual interaction with the system will therefore also be a combined effort. But with the issue of measurements, and later with the matter of the frequency of meetings, a new question is asked for the first time: "Who's going to do it?" Who is responsible for pursuing the course of the workflow and transferring the information gleaned into the Excel spreadsheet? Who will facilitate the meeting with the stakeholders at regular intervals? Unlike methods such as Scrum, there is no role in Kanban that is comparable to that of Scrum Master. It is also not stipulated that the team leader or department head should take over these tasks.

The answer, which for some is unsatisfactory despite it being the only correct one, is this: it depends on what the team can agree on. In many organizations, there is one person who feels particularly responsible for the Kanban change and is prepared to take on these tasks. Most of the time, these particular employees spend their whole day with Kanban and coordinate several teams simultaneously. Similarly, it can also be the case that the team itself doesn't only want to be in charge of the construction of its Kanban system but in fact also wants to be responsible for all aspects of continuous operation. Often, the solution turns out to be a rotating duty: each employee spends a week in charge of noting down the measurements and facilitating the discussions. There are no magic formulas for the responsibilities, or rather the sense of responsibility, but rather only initiative that must be captured.

Susanne Schweizer's development team initially seemed to have made a slightly less happy decision concerning measurements. Full of get-up-and-go, her team got derailed tracking bugs in the production system. "Customer satisfaction" was, of course, the goal. Error-free production systems certainly have something to do with that. However, there was one thing that the group of developers and testers hadn't considered: this also includes bugs such as pixel errors in individual graphics. These lie outside the perceptual range of the customers but are nevertheless termed bugs in the metric. On top of this, the metric also caused developers and product managers to get tied down in long conversations: was it really a bug or were the product managers defining the term "bug" differently than the developers? Thankfully, this discussion took place in one of the first retrospectives on the subject. Together with her team, Susanne Schweizer reflected once more about the errors that should be measured and those that shouldn't. Following this, the team and the product managers sat down around a table in order to create a shared understanding of errors.

Hadn't Ms. Schweizer observed the operational running of the Kanban system, lots of energy would have been unnecessarily wasted on a finger-pointing culture, simply to steer the measurements in the desired direction. And this is contrary to the philosophy of measurements, as well as that of Kanban. The eternal debate as to what is and is not allowed to appear in the measurements doesn't promote customer satisfaction, nor does it promote the development of a Kaizen culture.

17.5.1 Selecting the Appropriate Measurements

If measurements are already being discussed during the workshop, then the team has probably gathered a few metrics by this stage. But, as we've already said, not all of them will be necessary. Or perhaps the team is only now beginning to consider the issue. The workshop facilitator can make use of a simple approach here, regardless of the status quo, in order to establish the metrics that the team can use at least for the first few weeks to begin working without a hitch.

17.5.2 Step 1: Teamwork: Identification of Possible Measurements

- By now, the team has a Kanban system that is ready for use. Additionally, the results of the review workshop are hung up on the wall, just like before, to guide the team. During operations, the selected metrics must show whether the team has made any kind of progress toward the improvement wishes or whether they've solved the problems identified in the interviews. Facilitator and team take another good look at both these goals: the Kanban system they've developed and the results of the review—possibly also the notes regarding the measurement ideas that were generated during the workshop.
- The questions for the team are now as follows: What do you think? What measurements can help us at the beginning of operations? Which do you absolutely need in order to understand that you're on the right path, and how can these measurements be modeled? Again, this is an appropriate issue for small groups.
- The groups present their results in the plenum.

17.5.3 Step 2: Determining the Initial Handling

By the end of step 1, or perhaps even earlier, the team has agreed upon the measurements it wants to implement in the first stage of operations in order to be able to track the changes (at the beginning it's normally mostly about lead times). The team now knows what it wants to measure, but how is this best incorporated into their daily routine? At this point in our workshop, we provide an overview of the possibilities. The team itself must then reach consensus on how they want to organize the collection of their measurement data in their everyday work:

- **Visibility**: Measurements that are exclusively carried out in Excel spreadsheets or other programs are usually very easily forgotten, but the results of the work the Kanban team does should be no secret. On the one hand, there are stakeholders who have a legitimate interest in being informed about changes and the progress made by the team. On the other hand, an open representation is really an instrument for the team itself in order to keep everyone equally well informed. It's more laborious to constantly have to open a file than simply stand in front of a flipchart with a cup of coffee in your hand. People who spend their working hours busying themselves with technical problems or developing software are sometimes afraid to use the simplest solution. Nevertheless, the workshop facilitator must at this point inform the team that there is to be a policy of transparency regarding the measurements and that they are therefore to be displayed clearly and visibly in the room for all the team members, stakeholders, and other passersby to see.
- **Setting the measuring points**: If the team hasn't selected too many measurements, they can simply each be recorded on their own flipchart, by hand, and in *real time*. Remember Russell Ackoff: "It is better to use imprecise measures of what is wanted, rather than precise measures of what is not." The daily stand-up meeting is the ideal context for setting the daily measuring points once team members have completed a job. Of course, this doesn't prevent anyone from transferring the data over into an Excel spreadsheet—it just shouldn't be used as the only measuring instrument. This data should be the measurement archive of the team, and whenever a new flipchart is begun, the previous processes should be hung up in diagram form next to it in order to make the development clear.
- **Daily routine**: The suggestion that daily, real-time measurements should be made isn't always immediately accepted. Teams can sometimes see this additional task as an overhead, an additional burden. Before Kanban, most teams carry out measurements because they have been told to do so; only a very few carry out measurements because they recognize them as an instrument for their own further development. At this stage, the facilitator should make it clear that we're here talking about a matter of minutes every day—indeed, they can even be part of a meeting that's taking place anyway. The principle of a daily routine is essential and should be adhered to systematically: it is only when events, for example, blockers, resolutions regarding changed WiP limits, and systemic

changes such as a new employee's arrival, are noted in real time that a realistic representation of happenings and the changes they affect becomes recognizable in the analysis. Reconstructions based on memory are always rife with errors after a certain amount of time and demand a far larger expenditure of energy than noting down a few points every day. We once experienced a team that very diligently held onto all their tickets for analysis at a later stage. Daily events weren't noted down on them. After completion, all tickets landed in a drawer. They were "collected" there for 3 months after which a team member was ordered to evaluate the data. Of course, nobody could remember quite what had happened in the last weeks and at what time. The employee spent several days sorting through these tickets in order to generate precisely *no* information. For at least 3 months (plus a little time for the follow-up work), this team had absolutely no idea about their work process and the impact of their efforts to change.

- **Patience**: With so many changes and new daily routines, it's natural to want to know as soon as possible what it's all meant to achieve. In the context of Kanban metrics, patience is probably one of the hardest things to have. As we already mentioned at the start, it is only after a certain period of time that changes in the system become visible when translated into diagrams. Patience means refraining from interpreting every alteration you notice in the diagram following an event or a change. Measurements should be observed over a period of time in order to rule out singular phenomena.

By now, the team should have recognized that you are brimming with pragmatism as facilitator of this workshop. At the end of this section of the workshop, the general catchphrase should be: "Let's give it a go and then we can change it again!"

SUMMARY AGENDA POINT 5: DEFINING THE MEASUREMENTS

Marking out the framework

Our focus is on the utility of the measurements. We restrict ourselves to that which is significant, useful, and insightful in order to light the fuse.

Process

1. Identifying the best fitting measurements or measurement ideas that have already been collected as the goals.
2. Selecting measurements based on their informative strength: Which metrics best highlight the changes?
3. Determining responsibility for and administration of the measurements in the daily routine.

17.6 DETERMINING THE FREQUENCY OF MEETINGS

Strictly speaking, meetings are a waste because they don't deliver extra value to the customer. They generate coordination costs and don't always have the desired effect, since all too often the room is vacated following a discussion even though no concrete results may have been reached. Thus, it will come as no surprise that there may be some sighing and groaning in the room during the workshop when it comes to the subject of Kanban meetings. Yet another meeting, moreover a daily meeting, more coordination with the stakeholders, more coordination between the stakeholders.... It is highly likely that some of the team members begin to feel a little overwhelmed. Their reservations are understandable, since, more than any other instrument of organizational communication, meetings often end up being manifestations of self-representation and self-legitimization. Scott Adams, creator of the unforgettable Dilbert, says: "Now I understand that meetings are a type of performance art, with each actor taking on one of these challenging roles:

- master of the obvious;
- well-intentioned sadist;
- whining martyr;
- rambling man;
- sleeper [1].

Participation in meetings is an issue within corporate culture that, ironically, is software driven (especially in larger organizations). Outlook and Lotus Notes do not create a bond of participation but instead always leave open the possibility of indecision or even rejection of appointments. You can force others to meetings by grasping hold of their calendar directly, without having said a single, nonelectronic word to them. And these people can remove the appointments from their calendars just as easily, without reason and discussion.

We repeatedly notice extreme opposition in situations where Kanban systems have been developed for a team in the course of process engineering and simply enforced on them. Usually, we're then brought in because at some point or other nothing works anymore and the Kanban change initiative is stuck in the same dead end as all the other change initiatives that were created without dialog with those concerned. If, on the other hand, teams are given the opportunity to organize themselves and if they are allowed to develop their own path to the goal, then such things as meetings organized according to certain criteria no longer seem burdensome but indeed logical.

Meetings are a waste, but they are useful waste—waste that a team needs for coordination and improvement, just like a can of energy drink. Therefore, we accept these coordination costs, although they will of course be optimized as much as possible through efficient and effective modeling. The only two questions that still need to be asked are:

1. What meetings should take place?
2. How frequent should these meetings be?

In order to shape the workflow as optimally as possible, meetings—which, for example, determine the size of the input queue, are a means of creating trust between the team and the stakeholders in terms of releases, and, if we think about the preceding chapter on measurements, also support sensible measurements—should be held with a certain frequency.

TIP FOR MEETING MANAGEMENT

Once things are operational, the team will itself realize how individual meetings can be better target oriented and where the pitfalls of inefficient communication lie and be able to solve them accordingly. *As facilitator of the workshop, it is your job at this stage to put a particular emphasis on creating an awareness of the sense and utility of the individual meetings.* Agreement upon the nature and frequency of meetings doesn't happen because Kanban says it must. It should happen because the team sees the potential for further development. However, experience suggests that, regardless of anything else, teams should include the following meetings:

- Daily stand-up meeting
- Team retrospective
- Queue replenishment meeting

At the beginning of a Kanban initiative, operations reviews are seldom an issue and don't solely lie within the team's area of accountability. The decision regarding their implementation must be made at an organizational level because they involve the coordination of several Kanban teams. The release planning meeting is optional because some teams, for example, in support, simply don't need it.

In this workshop, it is important to emphasize that once meetings have been agreed upon, they *must* be attended by all team members and any stakeholders involved, even, or rather, especially if people are up to their eyeballs in work and the team is in emergency mode. Coordination and agreement are far more important in such situations than in quieter periods.

17.6.1 The Daily Stand-Up Meeting

Today, we live in a world with more communication possibilities than ever before. The problem is that we often use things in the wrong way and, intentionally or not, create distance and a lack of connection through the choice of medium rather than mutual understanding. It's almost tragic to see the extent to which e-mail excesses have replaced direct communication in organizations. Instead of sorting things out face-to-face in a couple of minutes, thereby establishing a binding responsibility for the next steps, everyone affected—even tangentially—is copied in. At the end of these disorientating back-and-forth discussions, no one really knows who's meant to do what. The result is that nobody does anything. And yet so much energy and time was invested in these utterly unproductive discussions. This is precisely what daily stand-up meetings are designed to prevent (see Chapter 6).

What Is the Purpose of the Daily Stand-Up Meeting?

- It provides the team with a real-time overview of the workflow and whether it is running obstruction-free or blockers exist.
- Every team member knows who's working on what, reducing the time spent making enquiries.
- Blockers, problems, and bottlenecks are discussed together, which means solutions can be found more quickly than when one person sits there alone, going crazy over the problem.
- Are the work items still within the stipulated service-level agreements or not? Through daily discussion, the team provides itself with the possibility of being proactive rather than putting off decisions and irritating the stakeholders.

Conditions for the Daily Stand-Up Meeting

- The daily stand-up meeting does justice to its name: it happens on a daily basis at the same time and in the same place, in front of the Kanban board. The deadlines for the whole year are established in the first meeting. The team chooses one person responsible for picking the dates and informing everyone about them. The opt-out reasons are holiday, sick leave, or being kicked out of the company.
- In the daily stand-up meeting's plenum, we only discuss what everyone needs to know. Conversations between two or three colleagues on a specific topic take place after the daily stand-up meeting in order not to unnecessarily disrupt the work time of others.
- The team agrees on who facilitates the meeting. Is it always the same person or will the responsibilities rotate regularly among all team members?

17.6.2 Team Retrospectives

Viewed very one-sidedly, retrospectives generate considerable coordination costs. The team concerns itself with nothing and no one other than itself. As experts, we regularly experience situations where teams have had no previous experience with the critical examination of their own work. After their first retrospective, these teams then say to us: "We may not have been doing our normal work but nevertheless that was the most productive time we've had in a long time." In a retrospective, or indeed the period of time immediately after one, something very significant happens: the team comes to realize that it is itself the core of all improvement work. The team has control over whether something changes for the better or not, because improvement can never be sought externally and can never be expected just to "happen." Improvement and change always begin with the individual, and this needs space in the work process. As we already said in Section 6.4, before there can be a culture of *continuous* improvement, a culture of improvement must first evolve. For teams that have never worked with them before, retrospectives are an important trigger for the change process.

What Is the Purpose of the Retrospective?

- In this meeting, the team observes its achievements over a given period of time. What worked? Why did it work? What didn't work, and why not?
- Are there particular problems such as bottlenecks or aspects of the collaborative work with stakeholders and between each other that the team should take a closer look at?
- What insights has the team gained through its measurements? What consequences do these insights have for the alignment of the system?

Conditions for the Retrospective

- The team comes to a fundamental agreement as to whether it wants to use retrospectives or not.
- If the decision is in favor of retrospectives, then we recommend beginning with a weekly frequency. As with the daily stand-up meeting, the deadlines are fixed and a team member takes over the communication of these deadlines during the first retrospective.

17.6.3 The Queue Replenishment Meeting

The queue replenishment meeting exists in order to make sure that work items reach the input queue in an ordered manner (see Chapter 6). In this meeting, the stakeholders agree on the order of the work items. The following are to attend the top-up meeting:

- At least one member of the team
- Everyone who has given the team work items
- Everyone who receives completed work items from the team
- Everyone who could contribute to a decision regarding what the team's next work items should be

Before a team begins thinking about the initial frequency of the meeting, it first needs to think carefully about who they want to participate in it.

17.6.3.1 Selecting the Participants for the Queue Replenishment Meeting The team's stakeholder maps evolve in the preparation phases before the workshop and during the identification of the work item types. The descriptions given by the stakeholders normally remain pretty general and aren't given precise names. The workshop facilitator brings these stakeholder maps out again and goes through the naming process with the team.

Before the exercise, the team should be given the helpful tip that although the more participants in a meeting, the longer decision-making might take, the decisions made by a greater number of people may well be of better quality. The goal shouldn't be to list all possible employees from all departments and all customers with whom the team has the remotest form of contact. What we're aiming for is the creation of a list of people who are in a position to make the decisions that affect the team's work.

This list is not final; it will again be reviewed when the Kanban system is brought into operation.

17.6.3.2 Establishing the Frequency of the Queue Replenishment Meeting As we now know, the frequency of the queue replenishment meeting depends on the team's throughput and the size of the input queue. But at the beginning of our project, both of these are very unstable and will probably change as soon as the Kanban system is brought into operation. Depending on the starting point, three different approaches present themselves:

1. **Establishing the frequency based on experience**. The team knows its territory very intimately and knows precisely what its lead times look like and perhaps even has records of them. In this case, the team members can usually form very good estimates of how often they need top-ups for the input queue. The frequency of the queue replenishment meeting is determined by consensus based on this experience.
2. **Feeling instead of frequency**. Some teams begin with the assumption that they do not know how long it takes individual tickets to be completed. Accordingly, they have no kind of indication of the throughput of their system. Additionally, the team members know that they currently are far in excess of the desired WiP limits. In such situations, we usually ask the team how many work items each team member is currently working on. Let us suppose we are dealing with a team of five people that wants to achieve a capacity of two work items per person. This would give a WiP limit of 10. In actual fact however, each team member works on seven work items simultaneously, which means a current WiP of 35. Therefore, there are 25 work items that first need to be removed from the system. A queue replenishment meeting is the last thing the team needs at the moment, because it's going to be occupied for quite a while with these work items. However, the working-off phase can be used to develop a feeling for the required top-up frequency. In the coming weeks, the team therefore has the task of observing and recording how quickly it completes work items. When everything is brought into operation, the stakeholders will be informed of the current situation and be given the opportunity to remove tickets from the input queue so that the team can get into its desired rhythm more quickly. Put simply, this means that for the time being there is no queue replenishment meeting but also that nobody should bring new work items to the team through the usual routes. The actual establishment of the frequency is put off until the team has made significant progress toward the planned WiP limits.

3. **Learning process**. If a team doesn't need to begin with completing old work items but also lacks informative records or reminders of previous lead times, pragmatism is once again called for. The team simply decides that the queue replenishment meeting will take place every Monday and that the WiP limit for the input queue is 10. It can turn out that this first stab in the dark fits perfectly, but it can just as easily turn out to be way off target. This isn't important, since the team now has two dimensions with which it can start and which it will realign if they're not a perfect fit.

17.6.4 The Release Planning Meeting

As has already been mentioned, a release planning meeting is of absolutely no interest in some areas. Teams that work with continuous deployment have their own specific release process, which cannot always be squeezed into a particular frequency. If such teams use Kanban, then they should continue to pursue their previous release approaches and should not be artificially forced into new molds.

But what is it like for the "normal cases"? Is it then possible or even necessary to determine a frequency for releases in the system design workshop? In Section 6.3, we spoke about the costs that arise during delivery: these are mainly coordination costs and transaction costs. Alongside the question of costs, releases are also a market-strategic decision. Sometimes, releases need to be planned according to the marketing perspective. The extent and frequency of releases differ markedly from organization to organization and only rarely are they something that evolves completely by chance. Usually, specific thoughts lurk behind them—thoughts that shouldn't simply be immediately thrown overboard. In this situation,

SUMMARY AGENDA POINT 6: DETERMINING THE FREQUENCY OF MEETINGS

Laying out the framework

If meetings are decided upon, then the team is making a commitment to participate in these meetings.

Program points

1. The workshop facilitator emphasizes the meaning, purpose, and conditions of the meetings.
2. Together with the team, decisions are made about when meetings should take place and with what frequency:
   a. Using the stakeholder map, it will be decided who should participate in the queue replenishment meeting.
   b. The team chooses the frequency of the queue replenishment meeting based on either experience, a qualified estimation, or simply through a learning process.

then, the team needs significantly more input in defining the frequency of meetings from the relevant stakeholders in the process of bringing the system into operation. What the team is able to say on the subject of the release planning meeting's frequency during the system design workshop is nothing more than transitory wishful thinking.

As facilitators, we would in this case normally advise teams to stick with the current release for the time being. To start straight away with a new frequency would be too much of a change. We absolutely insist on you trying to bring regularity to the deliveries—just not in conformity to a scheme that we define in this workshop. In the best-case scenario, the frequency would be the result of the interplay of all the other system elements to which we should pay more attention.

17.7 CONCLUDING THE SYSTEM DESIGN WORKSHOP

The team has now completed the construction of its initial Kanban system. During the meetings that took place in the final part of the workshop, issues relating to operations were touched upon and partially settled, although some matters still need to be agreed on with the stakeholders. At points in the workshop where it was sensible and indeed possible, the team carried out initial simulations with the system. However, up until now, they were only able to examine sections of the system; sometimes, it perhaps turned out that a simulation was only possible once all constituent parts of the system had been determined and were present. With the interim simulations, it was far more important that the participants better understood the concept and functionality associated with a specific Kanban element and that they were able to decide for themselves how this element should be adapted in order to better suit the team's work. The main focus therefore was the verification of what was being compiled.

Now, however, it is time for the team to experience for the first time what Kanban "feels" like at full capacity.

17.7.1 Simulation of the Entire Kanban System

If a team has only carried out a few simulations in the previous phases of the workshop—or indeed, none at all—then we are all for sending it for an imaginary 2-week or 1-month expedition to Kanban land. What's important here is the integration of the everyday and its usual and unusual happenings into the Kanban system and to experience how the team's work functions in this system. So what might happen during this expedition?

- As stipulated, the team begins the simulation with the queue replenishment meeting, considers which work items could be sorted in the input queue, and fills the relevant column on the board with the corresponding tickets.
- The first daily stand-up meeting: What is to be discussed here and how does this meeting proceed?
- Tickets are pulled across the board and blockers suddenly appear. How do we deal with this and what effect does it have on the rest of the work items in the system?

- Colleagues don't stick to the established procedure of the top-up meeting and instead create white noise. How do we make this visible and how do we set limits for it?
- An expedited ticket comes into the system. What are our policies for this? How do we react?
- Do our work steps take place in the order that we established or do some work items follow different sequences?
- It is Friday. What does our retrospective look like?

In summary, we take all the issues that have been discussed in the course of the workshop and run through them together, as one. Beforehand, the team should take one more look at the results of the review and build the problems addressed in the interviews into the simulation (e.g., ongoing disruption due to white noise). And of course, the team members are to implement all the policies they have documented in the course of the workshop.

Allow at least an hour for the simulation!

17.7.2 Conclusion

The team has performed excellently for 1 or more days and invested all its energy in the success of the Kanban initiative. Each and every individual has helped to make the team's work methods visible and they have therefore created a basis for improvement and change. Everyone has asked challenging questions as to what does and what doesn't function and the reasons for the latter. Many questions could probably be answered during the workshop; others might still be open and only answerable once the system is really up and running.

During the conclusion of the system design workshop, all participants should be able to say how they feel after this not insignificant change. Thus, the team members' feelings will become clearer. Are they full of enthusiasm, hunger for change, or have they succumbed to skepticism? Is there perhaps an opinion leader who is negatively influencing the team's mood? All team members should speak their mind. This feedback is important for the workshop leader because it will help to identify the areas in which more work needs to be done. Fundamentally, this workshop was the first giant retrospective, delivering input for the next developmental step.

WHAT YOU CAN TAKE AWAY FROM THIS CHAPTER

In the system design workshop, the team develops the Kanban system that it will begin working with and that will undergo constant alteration with increasing experience and growing insight. Through group work and discussions in the plenum:

- The boundaries of the Kanban system are established.
- The work item types to be processed by the Kanban team are identified.

- The work steps through which the work items pass are defined.
- The capacity of the system and consequently the WiP limits are determined.
- The classes of service are defined based on the costs of delay.
- The measurements that reflect progress made toward the goals are established.
- The frequency of the meetings needed for operation of the Kanban system is agreed upon.

Establishing the boundaries of the Kanban system and identifying the work item types:

- From the stakeholders identified in the diagnosis phase, we select those who have direct contact with the Kanban team. This "revised" stakeholder map leads to the question: "Which work items do we receive from the stakeholders, and which work items do we pass on to them?"
- The similarities between work item types are the basis for intelligent clustering (e.g., according to type, origin, size, or frequency of the work items). The deciding factor is the answer to this question: "What does the grouping of the work item types need to show us so that we can immediately recognize how to handle an incoming job?"

Identifying the processes:

- In this section of the workshop, we don't *define* the desired process; we *depict* the current, existing process.
- For every work item type, we examine the work steps it usually passes through. The degree of detail with which the process is depicted depends on the environment in which the team is operating.
- The first simulation brings insight into whether work steps or work item types need to be aggregated or introduced. The Kanban board is worked on as long as it takes for the current process to be reflected accurately and until it is ready to be worked with.

Determining the WiP limits:

- On the basis of records or experience, we identify what percentage of the team's work time is spent on each of the individual work item types. The capacities are then converted into WiP limits based on the number of employees. The WiP limits are then distributed across the individual process steps.
- Buffers and queues are not included in the distribution of capacity, but instead should be limited so that they don't become a hidden car park for work items. The size of the input queue is initially based on estimations and will be recalculated once the system is up and running.

- If the team has gained experience with the Kanban system, then the focus when setting the WiP limits is on the workflow and no longer on the number of team members.

Determining the classes of service:

- Initially, the goal should be to focus on fewer classes of service. The premise is: don't depict the exceptions, depict the rule!
- Each class of service is accompanied with team-specific policies regarding handling and capacities.
- The definition of the classes of service can lead to intense discussions about the current problems. It is important to allow these discussions to take place because the causes can very often be identified here and the problems resolved.

Defining the measurements:

- The performance measurements for a Kanban system should be as simple as possible and should support the attainment of the stipulated goals.
- The right measurements show a team where optimizations are to be made.
- In the workshop, the team decides who is responsible for the regular recording and evaluation of the measurements.

Frequency of meetings

- Meetings are important for coordination and improvement. In the workshop, the team clarifies which meetings should take place and how frequently.
- The frequency of the meetings is crucial for an optimal workflow: it determines the size of the input queue, supports informative measurements, and is a means for creating trust between the team and the stakeholders.
- It is important to create awareness of the fact that the meetings agreed upon are not optional: everyone *must* attend.

18

OPERATION

In "The Jungle Book of Leadership," organizational consultant Ruth Seliger writes that leadership is a bit like housework [1]; it usually goes on in the background and is normally only noticed when it doesn't happen. All of a sudden, dirty plates are left lying around, dust collects in corners, and the pile of dirty laundry grows. It's immediately clear that the invisible hand that usually takes care of everything is missing. Without attracting any attention, this invisible hand ensures the quality of our environment.

The same is true of a Kanban system's working environment. An invisible hand is also necessary to maintain the quality of operations on a day-to-day basis. The system's design must be kept under observation and if necessary adjusted. Bottlenecks must be identified and blockers eliminated. In other words, leadership is required.

The leadership of a Kanban operation is also for the most part invisible. Leadership is not a single project but a continuous process. It is a series of smaller and larger activities that are taken for granted although they require that someone does them.

The consequence is that it is quite easy to overlook the necessity of active operational leadership. The idea that everything just works on its own is one of the great myths of Kanban evolutionary change sustained by four potentially fatal assumptions:

1. **Kanban doesn't need professional change management** because the introduction of process visualization, WiP limits, and classes of service guarantees sufficient change.

Kanban Change Leadership: Creating a Culture of Continuous Improvement,
First Edition. Klaus Leopold and Siegfried Kaltenecker.

2. **Kanban automatically produces a sustainable cultural change**, that is, nothing further needs to be done in order for Kaizen to become an organizational reality.
3. **The creation of the system design is the crux of the matter** and from this point on improvements will just happen of their own accord.
4. **Kaizen is a logical result** of a Kanban operation.

What we have already established in connection with fundamental cultural assumptions in general is also pertinent to these Kanban myths: they are the aggregation of relevant learning experience, deeply rooted and for the large part subconscious. More than anywhere else, these myths seem to be embedded in the professional culture of technicians and process engineers. In this field, it is particularly tempting to reduce Kanban to a technical system: the organization is perceived as a machinelike entity whose change doesn't really demand anything more than the use of correct mechanisms.

If we take a quick look at the practical side of things, the social element of organizational and Kanban systems becomes clear. These systems require a combination of structure and emotion, process regulation and good communication flow, and principle and improvisation. For this reason then, neither the introduction nor the operation of a Kanban system follows the logic of a trivial machine. Instead, all stakeholders must be taken into consideration in order to be able to compile the real complexity of the value-creation process. It needs communication partners who are capable of critical examination and are prepared to begin this critical examination with themselves. And it needs modeling elements with which the available improvement potential can be used effectively. In other words, it needs **Kanban leadership**.

Practice also shows that there are key points at which the absence of leadership becomes noticeable particularly quickly. Returning to the housework analogy at the beginning of the chapter, these are points at which, in just the blink of an eye, particularly large amounts of dirty dishes pile up, dust gathers, and dirty laundry gets out of control. In Kanban jargon, these are the moments when significant bottlenecks become noticeable. In our experience, this happens most of all when dealing with the following:

- **A culture of failure and slack**, both of which play a large role in Kaizen and continuous learning
- **Professional facilitation**, which helps to shape improvement-oriented communication as much as possible
- **Resistance and conflict**, which can continue to appear during operations
- **The Kanban fire you want to carry onward**, in order to promote the positive development of your entire organization

In this chapter, we will once more look at our case studies in order to describe dealing with these bottlenecks with as much practical insight as possible.

18.1 MOVING FROM A FAILURE CULTURE TO A LEARNING CULTURE

Admit it: one of the reasons you're reading this book is so you don't fail in your Kanban initiative. That's completely ok, but don't forget that Kanban isn't a system for avoiding mistakes. Quite the opposite: your team will probably experience very intense phases where they notice one mistake after another. You might find that certain illusions suddenly lie in ruins all around your team. This will be a challenging time during which the desire for permanent change is put to the test. It might also be a time in which accusations are thrown across the room, where perhaps the entire initiative is questioned because too many worms are coming out of the woodwork—things that nobody really wanted to admit were there.

This might be the right moment to tell you that "evolutionary" is by no means a synonym for "easy." *Kanban is a system for not making the same mistake twice. Kanban is a system of learning.*

Mistakes are a big thing in our society. Go and peruse the advice section of a book shop or type "mistake" into the online archives of any business magazine: you will find plenty of articles about how to avoid mistakes, what the biggest three mistakes are, what the five cardinal mistakes are, what you must never do, what you absolutely must do, how to become perfect, or how to avoid any trap. Did you know that Christopher Columbus actually made a navigational mistake? He didn't stumble across India; he found a continent that couldn't have possibly existed. Ah, well, those Italians never do anything right, do they? When they're not busy striking, they go and discover a land that Queen Isabella of Spain hadn't reckoned with. She just wanted (someone else) to find the quickest route for herbs and precious stones from India, but it turned out to be one of the greatest discoveries in the history of mankind! And it wouldn't have happened if Columbus had taken the well-known, traditional route around Africa instead of an unknown route west. Many inventions are the results of "mistakes," simply because someone did something differently to how it was "supposed to be done." In Chapter 7, we met the watchmaker John Harrison who solved the nautical problem of longitudinal navigation. He had to endure 40 years of error and confusion, times of little support but many mistakes, before he finally developed the groundbreaking clocks H4 and H5 [2]. This navigational problem would have remained unsolved for many more decades had Harrison worried about mistakes rather than simply seeing what he could learn from them. One of the most important tools in Kanban, the Post-it note, evolved due to a type of glue that was initially treated by 3M's product department as an aberration and of no interest at all in terms of future utility.

In our social systems, we are trained to forget the ability to learn from mistakes as soon as possible. As soon as we are in school, we're forced to learn things off by heart. This is sensible up to a point, since reading, writing, and mental arithmetic are tools we ought to have a good command of. But we also learn that mistakes are to be minimized and that what is desirable is conformity. If we go on to study at university, we will have spent a good 20 years of our lives learning things by rote in order to avoid making mistakes, preceded by a maximum of 6 years of instinctive learning through trial and error. Society's concept of learning is effectively nothing more than the automatic reproduction of knowledge. Once we are released into the world of work, the pressure only increases because suddenly prestige, higher salaries, and social status are associated with not making mistakes. But fear of failure can cause personal crises. And if we can't function, then we have no value. And it drives organizations into economic crises, because employees who don't want to think and learn actively don't want to break new ground either. What is demanded of us is that we are experts, not learners. But it is often precisely this fear of failure that is responsible for us falling into the trap.

ACTIVE THINKING INSTEAD OF BEING TOLD WHAT TO DO: A WAY OUT OF THE MISTAKE TRAP

"I was in Southern Styria over the weekend," Head of IT Josef Drechsler told us while we were visiting one Monday morning in order to clarify some of the operational aspects of his Kanban system. "Do you know Gleinstätten? They've implemented a kind of "shared space concept" on the street. Have you heard about it?" We had of course read about it in the newspapers, because this approach had been pretty revolutionary in an overly regulated country like Austria. Traffic is a perfect example of how the fear of making mistakes can cripple a system. For every possible situation, there is a regulation that you can invoke in order to pass the blame—for example, for an accident—onto the other person. Instead of driving cautiously and simply giving way out of politeness or even because it makes sense, most drivers, cyclists, and pedestrians simply insist doggedly on their rights. But that's precisely what causes accidents. And on top of this, there's the wealth of road markings, traffic lights, and traffic signs that tell us what we are allowed and not allowed to do, thus preventing us from paying attention to the actual situation.

"In Gleinstätten they've removed all traffic signs along a 400 m stretch of the main road through the town, redone the road surface, and left it free of all road markings," said Josef Drechsler, explaining this traffic project. "I really cut my speed because I had to be aware of my surroundings. And then there were a couple of pedestrians who wanted to cross the road. Although it was really unusual not having any signs, the traffic seemed really quiet, flowed well, and people were almost being polite!" That made us even more curious. The "shared space concept" comes from Holland and is already in use in over 100 versions across the country. The fundamental principle is the equal right of all users of this public space. All "stakeholders" in the traffic system negotiate the best possible flow. Traffic signs and road markings are removed and in their place people need to be considerate, think about the other users of the space, be alert, and communicate. Thus, the road becomes a meeting space in which people are required to make *decisions* based on the actual situation, instead of simply doing what they're told to do.

The proponents of the shared space concept also emphasize that it's about an evolutionary learning process that is influenced by local mentalities and idiosyncrasies of the regional culture, although its essential nature remains unchanged (www.sharedspace.at). According to them, the important thing is to put the inhabitants of a village or town at the center of this kind of initiative, instead of simply imposing the concept on them. "And there I had to think of our Kanban system," said Mr. Drechsler, smiling. "The interesting thing about the shared space concept is that people drive slower, which leads to traffic flowing at a more constant, slower rate, meaning that people arrive at their destination quicker. In heavily regulated situations, the high variation in speed always generates queues. Oh, and with the shared space concept there are barely ever any accidents."

We mustn't forget that in the world of knowledge work we're part of a relatively young discipline. It's still firmly lodged in many managers' belief systems that the more you standardize processes, the fewer mistakes there'll be. This is certainly the case in industrial assembly. Here, people are dealing with (sometimes very) technically complicated products: a thoroughly organized process of movements and mechanical optimization is expedient and reduces lead times in their production. The difference is that with knowledge work we are dealing with *complexity*. On the one hand, the complexity of the subject matter: software development produces *new developments* again and again, and in increasingly short time periods. Certain elements can of course be reproduced, but in most cases, new problems need to be solved or new routes that were previously inaccessible made available. It's about innovation. And on the other hand, it is complicated at a social level, since software developers also want to express their creativity and intelligence. Standardization is a fast track to stubbornness in knowledge work, which is precisely what we don't want if we're trying to create continuous improvement. *Kanban is a complex, adaptive system that only works with the proviso that people see mistakes that have come to light as new potential.* Therefore, policies should be changed if they cease to be appropriate.

18.1.1 Yes, I Make Mistakes

But if we've all been calibrated in such a way as to avoid mistakes as much as possible, how can we find a way out of this trap? How can we create a learning culture? Josef Drechsler is by nature a very ready-for-change guy. However, his experience with the shared space concept gave him cause to think a little more about the necessary preconditions for the evolution of a learning culture in his organization. "Simply prescribing a learning culture is as pointless as prescribing change," Mr. Drechsler says, thinking aloud. "A learning culture needs to grow and for this to happen someone needs to take the first step. And I probably belong to the group of people in the organization that need to take that first step." The head of IT was right about that. As children, how did we learn what values are and how we should behave with other people? Our parents and those around us demonstrated these things in their own way of living. It's no different with adults. And you'll have seen this often enough in the world around you. If someone starts something new (think of the pull factor of Forrest Gump when he's running), some others join in. If the change to a learning culture is important to you, then there's really only one option: *begin with yourself—lead by example!* Make an example of your ability to make mistakes:

- At the **personal level**—"I have made a mistake."
- At the **Kanban system level**—"It's not working so well at this point. What can we do in order to avoid making this mistake in the future?"

But this doesn't mean, "Lead by example and deliver the solution." Just as with demonstrating your ability to make mistakes, refraining from offering solutions takes practice. At the beginning, Josef Drechsler had some trouble with this because simply standing there waiting for the others isn't really one of his strengths. But he was totally aware that the trick with the development of a learning culture is to have the team think actively. Simply impatiently announcing a solution would catapult him and his team right back into the culture of parameters. Now, whenever he feels like he already knows the solution to a problem and really just wants to shout it out while the team is still thinking about it, Josef Drechsler bites his tongue. He allows the team time and puts forward the question, "How can we make it better?"

Take small steps. Asking, "How can we double our productivity?" would simply overwhelm the team, due to its multidimensionality. In all probability, you'd probably just get answers like, "We need more people," anyway. Be as specific as possible and ask, for example, "How can we avoid constant spelling mistakes in our user interface?" Here, the team can suggest a solution, leading perhaps to a change in the process that could mean that someone will proofread the interface prior to release in the future.

The crucial point for the change from a failure culture to a learning culture is trust. If you really want your Kanban system to be successful, you should also bid the leadership model of command and control farewell. Trust the people with whom you work to come up with suggestions for solutions. This requires that you overcome the desire to have everything under your control. It might even require you to overcome the opinion that "If I don't control them, they'll just mess around." Yes, it might well be the case that at the beginning the team makes mistakes here and there. *But this is precisely the crux of a learning culture: being allowed to make mistakes but also learning from those mistakes.*

Trust is also not a one-way system. In some organizations, the trust is so deep that programmers are allowed to decide for themselves if and indeed when they upload something to the online live system. Thousands, even millions, of users are hooked in and there's no way a mistake will go unnoticed. And nevertheless, the programmers are trusted with this. Because trust doesn't just come for free: it entails lots of responsibility. Every employee has a personal interest in the software being delivered at the highest quality. If something goes wrong, they cannot shift the blame onto someone else.

18.1.2 A Particular Mistake: Slack

The human brain constantly tries to complete things that are incomplete. If we perceive something as incomplete, we think up the missing piece. We can read texts with gaps in them without a problem because our brain simply doesn't like these gaps and immediately draws on its wealth of learning and experience. Perhaps this is from where the desire to constantly use our capacities at 100% comes. The old way of thinking is that employees working at 100% are efficient and generate profit. In business terms, an employee working at 80% is a sinking ship. It's a simple mathematical calculation. So if management sees someone staring into space for too long, this will be considered an error in the system. In an instant, they'll be given plenty to do and have their remaining 20% shoved into another project so that everything adds up to that nice round 100%. But is this sensible?

Again, we stumble across the mixture of thinking patterns from the world of industrial assembly and the particular realities of knowledge work. People are easily replaceable in assembly line work: the movements don't vary and are highly standardized. If someone is off sick, you can replace them with someone else at the cost of a brief training period—someone who actually works at another location on the line but is currently free. In knowledge work, it's different. To be ordered over to a different project "just for a while" means getting acquainted with a whole new environment, familiarizing yourself with lots of new relationships, and getting to know the new ambit of responsibilities. Just this acclimatization effort uses up a large portion of time, especially if you're constantly being switched from one project to another. And on top of this, there's the fact that the employee isn't involved in the project at a 100% level but rather at 20%. This means that he'll be saying, "Sorry, you'll have to wait," when the main project is once again taking up his time.

Drawing on our analogy with road traffic once again, these "jumpers" in knowledge work create phantom jams. A traffic jam means that the system is running at 100% capacity. If there hasn't been an accident, traffic jams evolve primarily for two reasons:

1. **Due to overload**. More vehicles join the system than the stretch of road can permit.
2. **Due to variations in speed**. Just like Kanban, road traffic agencies' intention is to achieve a constant and regular flow of traffic. One way it tries to achieve this is with speed limits. However, we know that not all drivers travel at exactly the same speed, either because they can't or because they don't want to. This often has a butterfly effect (phantom jams). One driver changes lanes or brakes and triggers a sequence of reactions. Finally, a wave of braking is triggered and all vehicles slow down, sometimes even coming to a standstill.

Head of Development Susanne Schweizer knows this only too well. Ultimately, she was also constantly feeling the pressure to maximize her employees' capacity, right down to the last iota. Management was of the opinion that due to this it needed to relieve her of decisions via micromanagement, only to then complain when the projects once again took a little longer to be completed. "I should definitely write down this motorway analogy so I can bring it up at the next meeting," says Ms. Schweizer more to herself than to us, while taking some notes. "Looking at it this way, my people who are deployed first in one place then another are just like the drivers on a four-lane motorway: they constantly change lanes and then need to brake again and again because the other lane's traveling at a different speed. So they then change back, always left, right, left, right. And in so doing they slow down the workflow because they always need to fit themselves in once again." We didn't want to interrupt Ms. Schweizer in her thought process, but we nevertheless added that *Little's Law* is also pertinent (see Chapter 4). This law simply states that lead times are increased when new projects are constantly started. Maximization of capacities is thus not the same as high speed.

Knowledge creates variability or "slack," as we called it in Chapter 4, for the simple reason that not all people take the same amount of time to do the same thing. Some work quicker, and some slower; some receive the tricky part of a project, and some an easier part. Returning to an example from an earlier chapter: a business analysis establishes that a particular approach ought to be quicker by a couple of milliseconds. The developer works for 2 weeks on this. In knowledge work, we need to move away from the belief that only working at 100% capacity is good. One hundred percent capacity means a traffic jam and as a whole the system will become slower. In many organizations, it is simply not yet understood that slack offers three important opportunities:

1. **The opportunity for knowledge transfer**. Instead of making employees who aren't currently working at maximum capacity a hindrance on the Kanban motorway, they can become breakdown helpers and help remove a traffic jam. The first question that management and the employee should ask is this: "Do people downstream in the process need help? Is there a bottleneck there somewhere?" Assisting a downstream area can smoothen out the workflow again. More importantly however is that in doing so specializations are gradually shared. The employees learn something new; knowledge transfer and knowledge development take place; and another not inconsiderable advantage is that the mutual help strengthens the sense of togetherness in the team.
2. **The opportunity for improvement**. In the retrospectives, a Kanban team above all identifies what could be done differently or better and records it in the improvement backlog (see Chapter 8). This backlog isn't just so that people can act with a clear conscience: the work items should have a real chance of implementation. Slack makes this possible.
3. **The opportunity for innovation**. By now, there are plenty of examples of how organizations intentionally provide their employees with slack. A certain portion of the working hours may be spent on employees' own projects, plans, and ideas. Of course, not totally unrelated to what the organization does, since it would want to use the innovation that evolves as a result. But fundamentally there's nothing reprehensible about this, as long as both sides are clear about the conditions and satisfied with the setup. In most organizations, however, it's

> still somewhat different. Employees are made to work at 100% capacity and, rather than using slack, additional new programs are initiated for innovation, innovation teams are set up, and external consultants are hired—economical nonsense, sustained by the obsession with 100% capacity.

What actually needs to fundamentally change is the understanding of sensible and senseless uses of capacities. The difficulty is that it is not immediately understandable that running at 100% capacity in knowledge work doesn't mean that things will be done any quicker. Indeed, the opposite is most often the case: the process gets slower. A lower use of capacities actually speeds up the process. Therefore, a team needs a certain amount of slack in order to be able to be fast. This can be mathematically proved with *Little's Law* in Chapter 4. But you can also express it in another way: the only way slack would not evolve is if we were constantly to begin new work items (as is usually the case) or if all people were to be clocked the same, to operate the same, and to think the same. But then we're just running on the spot, which is not the goal of continuous improvement.

18.2 FACILITATION

German systems theorist Dirk Baecker writes that "The organizations of the next society will find out that what's necessary is to deploy [employees] in any area where things are precarious" [3]. With precisely the thinking that we advocate for Kanban leadership, he identifies four areas that in his opinion are particularly precarious:

1. **Leadership**
2. **Contact with customers**
3. **Organizational design**
4. **Negotiations with network partners**

Through these areas, Baecker predicts, the organizations of the future will discover that individual employees are not only to be used as functionaries viewed in terms of purpose and as the means to an end. They should be treated "according to their ability to relate perception and communication to each other" [3, p. 22].

How can this be achieved? How do you sensibly relate perception and communication to each other? One pragmatic answer to this question is quite obvious: through coordinated communication. In complex environments, this basically means: through professionally facilitated meetings. However, these days, as the American consultants Sandra Janoff and Marvin Weisbord have observed, meetings take place as frequently as particles of dust settle, and they're just about as popular [4]. "Time wasters," "burdensome duties," "one-way communication," and "meetingitis" are just some of the classic complaints related to unproductive discussions. As the saying goes, meetings are where lots of people gather and nothing is achieved.

These days, Helga Rösner, Head of Development for a media organization, knows that this doesn't necessarily need to be the case. As a matter of fact, Ms. Rösner had always thought of herself as a good manager and one who was always on top of her meetings. The gathering criticism of her leadership style, particularly of what one of her colleagues once described as "concerted commanding," made Ms. Rösner insecure. Was the criticism well founded? Are there really "completely other possibilities for shaping communication," as one of her colleagues suggests? Or is it perhaps simply unavoidable that some daily stand-ups last a little longer, some retrospectives aren't always interesting for everyone, and the operations reviews are dominated by large amounts of information delivered by one person standing in front of the rest?

"Ultimately, it was more luck than a decision," Ms. Rösner told us when we asked her why she had finally opted for further training in the field of facilitation. "It was more of a spontaneous decision than something I'd planned for a long time," Helga

DON'T JUST DO SOMETHING, STAND THERE

In their groundbreaking book *Don't Just Do Something, Stand There*! Sandra Janoff and Marvin Weisbord created a practice-oriented guide for professional facilitation. In order to be at the cutting edge with your meeting design skills, Janoff and Weisbord believe that in every meeting, you need to:

- Bring the right people together, with the right focus, at the right time. This means establishing a good match between the participants, goals, and duration beforehand.
- Treat people as they are, not as you would like them to be.
- Not tackle change by looking at the behavior of others but by looking at your own behavior—the same goes for managing the conditions over which you as facilitator have control.
- Give everyone the possibility of expressing themselves and politely paying attention to each other.
- Make sure that you as a facilitator hold back, both physically and verbally, as much as possible, that is, to set the stage and then very visibly hand things over to the participants.
- Use your own feelings as a feedback source for the group event while at the same time being aware of your own "hot buttons." Ultimately, your own fears shouldn't lead to actions that affect the whole group. In their own words:

> The more we learn to live with uncertainty and remain curious about what's to come, the better prepared we are to value each group's struggle. So we resist the tendency to manage our own anxiety by talking, asking questions, explaining, repeating, or changing the subject. When we're not sure what to do, we don't do anything [4, p. 172].

Rösner explains, "but turned out to be an absolute godsend." With "luck," Ms. Rösner refers to her participation in one of the facilitation training courses that Sandra Janoff and Marvin Weisbord offer. The "godsend" refers to what she took away from this course for her Kanban everyday operational routine.

Thanks to the training course, it has become clear to her how important a professional facilitation attitude is. In the following section, we bring together the defining features of this attitude.

Ms. Rösner uses this training course to brush up on her facilitation skills. Communication cannot be steered according to a plan, as Ms. Rösner confirms in a couple of noteworthy exercises. As the heart of social systems, communication remains unpredictable, subject to breakdown, and afflicted with the risk of talking at cross-purposes. However, it is possible to control the conditions of communication. The structural elements are:

- **Setting clear goals**. "What do we want to achieve in the queue replenishment meeting, for example?" asks Ms. Rösner, summing it up. But also: What isn't it about? What doesn't belong in this meeting?
- The **right participants**. Whom do we need in a retrospective? Who needs to be invited so that the status quo can be represented as well as possible? Whose perspectives can we not afford to leave out?
- A **structured agenda**. How detailed does an operations review need to be? Which design elements can be used to provide the most useful analysis possible?
- **Good time management**. What's the upper limit for the duration of a daily stand-up? When do I intervene in order to stick to the plan? What do I do with issues that obviously need more discussion time?
- **Focusing attention**. What means can I use to make sure that we concentrate on the most important points in a retrospective? How can I get everyone to participate as actively as possible? And finally, how do I compile the most important results of my meeting?

There's a wealth of tools for focusing attention at your disposal. As Ms. Rösner recalls, there is first and foremost a long list of questions [5]:

- **Opening** questions, like the many you've come across in the course of this book: Why Kanban? Who should be involved in the change initiative? How do we want to proceed? What's been working well recently? What hasn't been working so well? What WiP limits should we set ourselves?
- **Closing** questions. In a Kanban context, these might be the following: Is there a blocker here? Do we all see a bottleneck here? Do we need a new class of service? etc.
- **Scaling** questions, typical for a solution-orientated approach: How well have we met a certain criterion on a scale of 1–10? What would we need to do in order to move from value x to value $x+0.5$?

- **Circular** questions, particularly useful in evaluating stakeholder orientation: What do our customers think about this bottleneck? What do we think our boss would say about this idea? What do you think a Kanban expert would recommend at this point?
- **Paradoxical** questions that, particularly in the context of the Kanban retrospectives, provide new, often surprising perspectives: What would we need to do to keep everything as it is? How could we make the problem worse?

PARADOXICAL INTERVENTION: SEEING THE GOOD IN THE BAD

Till Eulenspiegel was a renowned master of social challenge. Indeed, he once showed how it is possible to achieve good by announcing something bad. One day, when he was in a hospital, he told all the patients about a magical cure. For this cure, he would only have to burn one of them down to a powder and with this powder heal all the rest. Suddenly, everyone was well again. How did Till Eulenspiegel achieve this? Well, he simply told them that he had been employed to sacrifice the sickest patient for the benefit of the rest.

Brenner, the central character in a work by Austrian author Wolf Haas, also philosophizes about the good in bad things:

> But it's always interesting how in life a setback often turns out to have been a gift from heaven in hindsight. When, later on, one must admit it and say, you see, if my house hadn't burned down to the ground, if I hadn't lost my well-paid job and my beautiful wife, then I probably never would have fished this highly interesting crossword out of the rubbish bin [6].

And finally, there's also the positive observation of the person who's just been pooped on by a bird: "What luck cows can't fly!"

It's clear to Ms. Rösner that she doesn't just need to ask questions, but must also listen to the answers. "If you ask questions, you run the risk of getting answers," is the advice from the training course that still echoes in Helga Rösner's mind. It's accompanied by the associated point that simple acknowledgment of these answers does not go far enough. Rather, Kanban leadership requires the willingness to engage in an active reaction to these answers. This can be achieved with the well-known techniques of active listening:

- **Paraphrasing**—repeating what has been said in your own words: "From what I've heard…," "What I'm hearing is that…"
- **Acknowledging**—bringing out the emotional message of your conversational partner: "I get the impression, you are frustrated about…" "You seem to me to be particularly worried that…" "I can sense your enthusiasm for…"
- **Reframing**—presenting what has already been heard in a new light by considering the "burdensome" interventions of a stakeholder, for example: "He seems to want to make this decision at all costs. It must be very important for his organizational unit."

- **Summarizing**—rounding up previous communication in order to secure midway results: "I have the feeling we can now say the following…"

For Ms. Rösner, securing results is particularly important in the context of the Kanban operation. She can use a variety of visualization techniques to secure results in meetings:

- **Card surveys** in which short statements are written on cards and clustered together
- **Brainstorming on sticky notes** in order to collect as many ideas as possible concerning a particular question as quickly as possible
- **Classic posters** on which the most important thoughts and observations are recorded
- **Mind maps** and other forms of visualization such as the **stakeholder map**
- **Organizational images** such as those used in the team constellation or those used through metaphor and analogy (the team as a zoo, football team, landscape, etc.)
- **Symbols** such as those Ms. Rösner once used at the opening of a retrospective in which all participants were invited to bring along a real object that symbolized the current working culture

Last but not least, Ms. Rösner has taken away a few tips on how she can encourage interaction in her Kanban meetings. Together with the previously mentioned structural elements including goals, participants, agenda, and time, she frequently makes use of the following elements from her facilitation training:

- **Go-arounds** in which every participant says a few words and must be listened to by everyone else before the discussion is allowed to start. It is similar to the so-called flashlight in which each person makes a short statement about a specific issue (e.g., "My statement concerning…"). Both strengthen shared opinion forming and give participants of a quieter nature the chance to put forward their position.
- **Buzz groups** are particularly useful in longer presentations, for instance, in the operations review, where it is recommendable to digest what's been said in smaller groups (three or four members) and possibly generate a brief commentary or question for the plenum.
- **Group work** for delving deeper into issues—in retrospectives, for example—often associated with a particular visualization exercise.
- **Dynamic presentations** that are not in the tiresome one-person-standing-at-the-front format but rather as a gallery, where all participants walk around and converse with each other as and when they want to; in so-called cross-cutting groups, composed of participants from previous groups who then present each other their respective results; or in the form of interviews in which the respective group is asked specific questions about the group results by chosen "reporters."

- **Dialogic forms of discussing results**, such as the so-called fishbowl, a circle of seats in the middle of the plenum in which delegates from the work groups can together compile their insights.

Since her facilitation seminar, quite a lot has happened in Ms. Rösner's area of work: through Kanban leadership, she has significantly advanced her understanding of Kaizen; the continuous improvement of leadership performance requires her to be a role model herself; encouraged by the positive reaction of her team, she has undertaken further improvement steps. For a long time now, it's been clear to her how much a culture of continuous improvement is determined by the quality of the communication that takes place. The chronic shortage of time and variety of tasks that are the bread and butter of knowledge work put particular demands on professional facilitation.

> *By now, Ms. Rösner can very well understand "why the ability to shape internal communication efficiently—constantly becoming more challenging and simultaneously more liable to breakdown—has become the decisive bottleneck in the system"* [7].

Ultimately, from her everyday experience, she knows that the special dynamic in the world of media—her world—necessarily entails a large amount of disruption. And she knows how she can effectively work on the communication bottleneck using professional facilitation.

18.3 CONFLICTS IN OPERATION

The territory of organizational conflict is a large kingdom, reaching up into the highest realms of strategy and down into the darkest depths of everyday squabbling. And, in just a few moments, it can mutate and take on a completely different form.

The amoeba-like form of conflict is mostly due to the dynamic of the emotions involved. As we have already shown, these emotions don't follow any linear or defined pattern: what drives person A crazy leaves person B completely unmoved; whatever has person C merely shrugging their shoulders in agreement triggers deep concern in person D; whatever has team A jumping to battle stations leaves team B indifferent.

In many organizations, conflict becomes an issue when a new process is introduced. Kanban is no exception. Although evolutionary change happens in small steps, emotions still come into play. The introduction of change is approached at a lower level, so that the personal change processes are for the most part less dramatic and the systemic J curve doesn't nosedive so significantly. But there are no guarantees. As we know, things frequently don't turn out as expected.

There is just as little guarantee that emotions in change processes are resolved once and for all through a professional introduction of Kanban. Emotions are characterized by their ability to repeatedly return from the grave when:

- An important **stakeholder doesn't stick to the Kanban-specific agreements**.
- The **team falls back into old habits** and lets the Kanban system slack.
- Collaboration is plagued by **dysfunctional behavior**.

In our tried and tested fashion, we want to use specific case studies to show how you can overcome such phenomena in a solution-oriented manner.

18.3.1 An Important Stakeholder Doesn't Stick to the Agreements

Let us take a closer look at Susanne Schweizer's Kanban operation. Ms. Schweizer is the Head of Development in a midsized pharmaceuticals company. She had made a strong start with her Kanban initiative and, after some initial difficulties, was able to convince her team and customers in sales and marketing that an evolutionary change management approach was right for them. She was able to disperse the skepticism some of her colleagues still harbored at the beginning through intensive communication. Using real improvement steps, she had also managed to convince the customers—who likewise weren't all initially enthusiastic—that Kanban deserved a chance.

However, what the Head of Development hadn't reckoned with was the behavior of the director, who is also the head of the family that owns the organization. Although "the senior" (as this combination of manager and owner was known within the organization) had obviously been involved in the contracting phase, was constantly informed about the Kanban initiative, and even fired the official starting cannon for the operational launch, he turned out to be a major problem once things were running. Instead of sticking to the service-level agreements, he continued to intervene directly with the development team in order to get his wishes fulfilled. One developer was even called up to discuss a web shop problem with the senior in his office. The senior also viewed WiP limits more as "nice to have" luxuries than as genuine restrictions on organizational processes.

These interventions didn't just damage the team's morale, but they sabotaged the overall change initiative for which Ms. Schweizer is responsible, also casting a shadow over her authority as a leader. After she had spent a while explaining away these interventions with a comment regarding the senior's well-known personality—generally considered open-minded but in certain areas the classic organizational patriarch—a certain resistance began to develop in Ms. Schweizer. When the senior once again tried to sidestep the policies, Ms. Schweizer asked for a word with him.

The aim of the conversation was simple: without disrespecting his privileged position, Ms. Schweizer had to nevertheless make clear to the senior the necessity of sticking to the Kanban agreements. In order to achieve this difficult goal, Ms. Schweizer followed the advice from the change dialog she had already successfully implemented during the diagnosis phase:

- After a brief introduction in which she summarized the reason for and aim of the conversation, Susanne Schweizer invited the senior to describe the way Kanban worked, as he understood it. Cue the expected words of praise for

Ms. Schweizer's strength of innovation and the jovial encouragement to "keep up the good work for our company."

- In the second stage, Ms. Schweizer let her boss know her own perspective. She thanked him for the praise and responded by mentioning how much she valued his openness and the way he had gotten involved with the Kanban initiative. But she also made it clear that the director's behavior was having a negative impact on the initiative. She voiced her understanding that sometimes certain demands are very urgent. But at the same time, she emphasized that these demands may not be squeezed into the Kanban system outside the defined agreements.
- In the third stage, she invited the senior to join her in devising a possible solution. Ms. Schweizer suggested implementing a specific "expedite class of service" for the director's demands. An appropriate WiP limit and the impact the new class of service would have on the overall workflow would then be considered at the next queue replenishment meeting. In response, the senior promised to respect the process policies in the future and participate in the meetings on a regular basis instead of simply skipping them. "Kanban is truly important for me," he reassured her once again. "I am convinced that it is the right path for our future." Additionally, he asked Ms. Schweizer to absolutely "come and have a word in my ear as soon as I start running wild all over the place again." "It's difficult to teach old dogs new tricks!" he said, winking at her.

18.3.2 The Team Relapses into Old Habits

Let us consider Stefan Bergmüller's situation in order to better understand how a team can regress during Kanban operations. As second-level support team leader at a major financial services provider, Mr. Bergmüller introduced Kanban with his back against the wall, so to speak—in other words, at the behest of his CIO who saw a promising possibility in Kanban and specifically asked him "to give positive change one more chance."

Mr. Bergmüller had honestly had enough change. In actual fact, he was ready to throw in the towel at the end of his personal retrospective, given how depressing his personal situation appeared: chronic overload, lack of appreciation, lack of understanding, and hopeless chaos.

Initially, Stefan Bergmüller was by no means convinced that Kanban could actually provide what his boss promised. But as a loyal colleague, he didn't want to turn his back on his boss or on the organization in which he'd put 10 years' service—and least of all on his team, whose spirit and willingness to see things through inspired hope in him, despite all.

Stefan Bergmüller was surprised during the introductory training course—which not only the team but also the CIO and many clients attended—for at least two reasons. Firstly, there was a good atmosphere, even though those who had often made his life hard were also present. And secondly, the principles that were introduced were definitely worth considering. The phase of deeper analysis, where he again came into personal contact with many clients, provided the next surprise. Although he had known most of them for years, he learned many new things about them and was really surprised about some of their perspectives.

After the feedback and the energetic system design workshop, the Kanban system was confidently put into operation. The first weeks were simply sensational. Mr. Bergmüller would never have thought it possible that such decisive change could come about in such a short period of time. At the flick of a switch, the team could work in peace and quiet, services could be rendered as planned, and the clients even praised the support team for its work. The hectic daily routine was a thing of the past and a most unusual sense of calm pervaded the organization.

So of course it was typical that precisely at this moment, when things were calm and working well, Mr. Bergmüller fell ill and had to stay home for a few weeks. Upon his return, he was in for a very unpleasant surprise. The old, hectic routine had moved back in. Just like before, everyone seemed to be rushing around with uncompleted work items, the telephone was ringing like mad, and e-mails with new prioritizations were constantly being received. Stefan Bergmüller was lost for words. What had happened?

On the one hand, Mr. Bergmüller could explain this crisis situation very pragmatically: two of his top people had also been away during his absence. Nevertheless, he couldn't explain why the rest of the team had completely lost sight of all Kanban principles. Clearly, they had panicked and slipped back into the traditional coping mechanisms. The Kanban board hung, unloved, on the wall and meetings hadn't taken place in 3 weeks. Nobody had felt the need to take on responsibility for the continuous improvement, meaning that old habits had surreptitiously taken hold again.

It was clear to Mr. Bergmüller that something needed to happen as soon as possible. It couldn't go on like this! He organized a special retrospective for the very next day. In contrast to what had been agreed, no review had been facilitated in the previous weeks. The atmosphere during the retrospective was as tense as he expected. Although Mr. Bergmüller explicitly made use of the "prime directive" in his opening and explained that he didn't want to blame anyone, guilt hung like a cloud in the room. "Yeah, ok, we knew we were meant to be working with the board," one of his colleagues said, "but because of all the cries for help from our clients, we barely knew which way was up." "Somehow we just let it slide," someone else confessed. "We simply had so much work that there was no time for meetings." "I'm sorry" was written on one of the cards that had been circulated for the team to write down what had gone wrong in the previous weeks.

Despite the general apologies and self-justifications, some very constructive conclusions were ultimately drawn from the collected cards. Several positive things could indeed be said. It was decided that the Kanban system would again be run according to plan and that people would take the necessary time for communication with each other as well as with the clients. Each team member chose an avatar for themselves and the next day these were hung up on the board.

Stefan Bergmüller also took away several lessons on leadership from the retrospective. During the euphoria at the beginning of Kanban operations, he had clearly lost sight of the fact that the improvement culture was still in an early formative phase. Instead of keeping an eye on the learning process, he had sort of assumed that from then on everything would just fall into place. He had also assumed that others would take over the responsibility of leadership in his absence. Now, the meaning of a continuous process of leadership learning was clearer than ever before—above all, the importance of not carrying entire responsibility for the operation of the Kanban system on his own shoulders. The suggestion of rotating the facilitation of the Kanban meetings was the first step in the right direction and, after the initial hesitation of the team members, was accepted and implemented.

18.3.3 Collaboration Is Plagued by Dysfunctional Behavior

Nothing made sense to Josef Drechsler anymore. As head of IT in an infrastructure organization, he had been used to quiet, structured work ever since the introduction of Kanban. Lots of external pressure could be channeled, while internal processes could be given a new foundational basis through the clever system design. However, trouble had broken out between the developers and the operations team. "That's all because you guys aren't doing your job," one team member said accusingly to another, who then countered, "Really sorry, but we're not yet at the point where we can start doing your work for you."

"It's incredible how deluded one can be," thought Mr. Drechsler, as he remembered his hope that Kanban would mean a final purge of all the old conflicts. During the introductory training, the significance of mutual support had indeed received as much unanimous head nodding as the importance of seeing Kanban as a cultural change initiative. But in the retrospective, a massive free-for-all was now breaking out as to who was responsible for what. During the ensuing battle of words, it was claimed that certain people "let other people take the blame for their own mistakes, rather than owning up to them."

After Mr. Drechsler had failed to comprehend the actual conflict, let alone work toward a shared solution, the retrospective ended in a highly embarrassed atmosphere. "What have I missed?" Mr. Drechsler asked, finally. "How does a fundamental willingness for mutual support fit with the squabbling that clearly isn't only affecting the retrospective? And what's the reason for this squabbling?"

"They've simply been passing the buck to each other," Josef Drechsler said that evening to his wife, when she wanted to know what his view of the conflict was. The conversation confirmed Mr. Drechsler's suspicion that the accusations were merely symptoms of deeper lying problems. "If you want to resolve the conflict, you must look for the triggers," said his wife. "That's going to be difficult because, in all probability, you're a part of the problem that you want to solve."

"Good point," Mr. Drechsler thought to himself lying in bed. "I can't simply see it as a team conflict and exclude myself." After a not particularly peaceful night, it was clear to Josef Drechsler that he couldn't resolve this conflict alone. His wife's observation that he himself was part of the problem echoed in his mind. It was of course about leadership and his personal competence, whose limits he'd been feeling painfully since the retrospective. "What if my own leadership capabilities are a bottleneck?" thought Mr. Drechsler, with growing unease.

"It is a sign of strength and self-confidence to recognize the limits of your own knowledge and know enough to enlist outside help," Mr. Drechsler thinks, reminding himself of a sentence he read somewhere [8].

And indeed, it doesn't feel like an admission of weakness that he ends up seeking the advice of an experienced coach. Even in the first meeting with this coach, he felt sure that he had finally begun to find a solution.

"If leadership really is an important bottleneck in our current Kanban system, then it is precisely the right decision to strengthen our project in this area," Mr. Drechsler heard himself reasoning at the end of the first meeting. Another reason for his relief was that with the help of the coach he had drawn up a clear strategy for proceeding. The following had become clear to him:

- He contributed more to the problem than he was comfortable with. The feedback from the coach alone was enough to make this clear. After Mr. Drechsler had described the conflict as "kindergarten," "signs of immaturity," and "a sideshow," the coach described these expressions as belonging in the same basket as those that Mr. Drechsler criticized in the other team members.
- Finger pointing, mistrust, and degradation all have a past: they are expressions of a broader corporate culture that had a negative impact in many areas.
- What the coach reported from other similar initiatives could well be the case—namely, that structural problems become personal and that useful solutions aren't to be found purely at the level of individual behavior.
- Things desperately need a fresh start. Other routes must be found in order to articulate the mistrust, anger, and disappointment without falling back into old patterns of behavior.
- A series of small, trust-building measures would create a new cultural framework for the upcoming analysis.

The insightful plan was followed by a similarly insightful implementation. Building on similar experiences with other cultural problems, the coach supported Mr. Drechsler's leadership through:

- A **step-by-step approach** in accordance with the principle of Kaizen
- **Discussing** the planned steps very **precisely**
- The **shared creation of fundamental information** for all employees
- **Detailed one-on-one conversations between the coach** and selected key players
- **Focused feedback on the most important issues**, which also enabled Mr. Drechsler to carry out a comparison between his own impressions and those of the coach
- The **facilitation of a joint workshop** to which all employees were invited

At the end of this "Fresh start" workshop, Mr. Drechsler wasn't the only person to be surprised by the results. They had managed to define some very specific improvement steps. As the coach had suspected, the exact clarification of the interfaces and the agreement on a coordinated, multiteam workflow to maximize customer value were some of the highlights of this workshop.

But for many colleagues, the most amazing thing was how these improvement topics had been fleshed out. In the lightning round that completed the workshop, many compliments on the "pleasant atmosphere," "good working environment," "respect," and "working for each other, instead of against each other" were made. This came as no surprise to the coach. Firstly, he'd already heard similarly positive key values in the course of the individual conversations. Secondly, the workshop had opened with the question, "What should we particularly watch out for today in order to make a difference for the better, as a group?" So it wasn't at all surprising that many of the key values mentioned had also come up in the preparatory conversations.

The progress of Team Drechsler's story shows that the differences achieved in the "Fresh start" workshop turned out to be game changing. Many of the decisions were implemented with lightning speed. This led to considerable relaxation in daily operations and was even noticed by several clients. The key values defined in the workshop stayed as a kind of reminder. As Josef Drechsler explained to his coach during the postworkshop discussion, people repeatedly referred to the "charter" established back then, especially if someone lost their temper. According to Mr. Drechsler, this created security in the entire department, along with the feeling of continuously being reminded how to work together—in other words, the culture of collaboration.

Last but not least, Mr. Drechsler had the impression that he himself had profited greatly from this concerted cultural change. Firstly, he was of course very relieved that the operation of the Kanban system was running significantly more smoothly than prior to the workshop. And secondly, it had become clear to him how much he was able to contribute to the positive development as a leader. "Ever since I began taking the time to put my own leadership on trial, I don't just learn for my own benefit," he explains, proudly. "I also become more and more of a role model for the culture that I would like to create with Kanban."

18.4 CARRYING THE KANBAN FIRE ONWARD

But what do you do if everything works smoothly? If your Kanban system works like a dream? If motivation and value creation are on the rise and all stakeholders are happy? Is the evolutionary change finally complete? And are you then finished with the continuous learning about leadership that one of our clients once described as a "true Sisyphus challenge"?

"What should I do when I'm at the top?" the Zen student had asked his master in the introductory story in our book, before climbing up the 21-rung ladder. "You can stand there," the master had explained. "You can enjoy the view. You can climb back down. Or you can continue to climb, without any rungs."

The aim of this book was to give you the courage to continue climbing, to progress your improvement work beyond the end of our change management rungs. To conclude, we would like to show you what such progress might look like using two final case studies.

The title of this final section actually catapults us straight into the middle of one of our case studies. "Carrying the fire onward" were actually words that came from Herbert Krakauer, the department head of an organization that specialized in security solutions bringing together expertise in hardware and software. With these words, Mr. Krakauer was describing his understanding of leadership, seeing it as a process that cannot remain static in one's own area of work. "There are lots of different elements in leadership," he said, explaining it to us. Paraphrasing him, it's about:

- **Strategic leadership** that sees Kanban as a useful means for adding value
- **Operational leadership** that finds inspiring solutions to problems as quickly as possible
- **Culture-oriented leadership** that also takes the well-being and the work satisfaction of all organizational members into consideration
- The leadership that resides in any kind of **good conversational leadership**
- The **allure** for others to take leadership in implementing Kanban for themselves

It's relatively easy for Herbert Krakauer to whip up a little forest fire of excitement. With 250 employees, the company is small enough to keep everyone in sight. Moreover, as a member of the board of directors, he himself is in a privileged position and able to focus people's perception and communication on a broader Kanbanization, so it comes as no surprise that he has by now won over two further areas of the organization for evolutionary change management.

As Thomas Müller's case shows, the fire can be spread in very large organizations too. As Head of Cards & Projects in an international energy organization, Mr. Müller finds plenty of opportunity to start fires. Pragmatically, he begins by telling people about tried and tested Kanban initiatives, namely, by word of mouth during organizational meetings, in the form of written reports based on selected metrics, and during interdepartmental exchanges with his colleagues in management.

Due to the reorganization in his area of responsibility, at the end of which a presentation on the new mission statements was made, all other stakeholders are already well informed. He keeps his clients in the marketing sector up to speed with regular operations reviews and also keeps them happy with the improved results. At one team's request, a Kanban board was even relocated from the office to the corridor in the interests of transparency. Anyone heading to the kitchen can consult the colorful Post-its and carefully created avatars. Day by day, colleagues in other areas can follow how these Post-its move across the board, see the team in action during their daily stand-up every morning, or observe small groups hard at work on a current problem.

Just in case, Thomas Müller has also arranged a comprehensive overview of Kanban. And, as it turns out, he is asked time and time again to explain this colorful operation.

KANBAN ELEVATOR SPEECH: WHAT YOU CAN SAY ABOUT KANBAN IN 30 S

Kanban is a method for the continuous improvement of your own area of work. You don't begin with a big change management project, but rather focus on a series of small change steps. You identify your most important business partners and together investigate the strengths and weaknesses in your current work processes. Based on a visualization of these processes, you use simple means to make things more efficient, improve your lead times, and create added value for your customers.

Of course, Thomas Müller had his best experiences as a Kanban fire starter in more informal surroundings: chilling out in the evening at a bar, during the breaks at official events, and at the half-business, half-private dinner parties with managers he's friends with. In these situations, you can do more than just strengthen trust and discuss positive or more challenging aspects in greater depth—in pleasant surroundings you can make the case for change work in a way that in other circumstances might come across as threatening.

It turns out Thomas Müller was onto a winner. While we were concluding this book, the Kanban fire was taken on by two other heads of department, while the service line lead had also taken part in a Kanban change introductory training course. Mr. Müller also agreed to start meeting up once a month with the other two heads to exchange their views informally. And finally, Thomas Müller was invited to a conference where he spoke about his practical experiences.

What can we conclude from this? Well, it shows us that active leadership can lead to far more than just good operation of your own Kanban system. It is true that evolutionary change management is based on a series of small blazes that you are constantly setting alight in your area of responsibility. But when considered as a whole, these small blazes can all together create a beacon of improvement.

WHAT YOU CAN TAKE AWAY FROM THIS CHAPTER

The assumption that after the operational launch of the Kanban system everything will run of its own accord is one of the great myths of improvement.

There is no automatic Kaizen system. In operations, active leadership is as important as ever for ensuring the continuity of improvements.

The key aspects of operational leadership include the following: the culture of failure and the correct use of slack, professional facilitation, dealing with resistances and conflicts, as well as carrying the Kanban fire within your organization.

In all areas, Kanban leadership is determined through an effective combination of behavior and technique. As a manager, you are constantly a role model for the attitudes and forms of behavior that you want to see around you.

Kaizen is sustained by your willingness to regularly put your own leadership on trial and make it the subject of continuous learning.

LIST OF FIGURES

Kanban Change Leadership: Creating a Culture of Continuous Improvement,
First Edition. Klaus Leopold and Siegfried Kaltenecker.

REFERENCES

PREFACE

1. Kaltenecker, S., Beyer, M., 2014, InfoQ: Kanban on track. Evolutionary change management at the Swiss Railways. Available at http://www.infoq.com/articles/kanban-on-track (accessed Oct 27, 2014).
2. Kaltenecker, S., Hundermark, P., 2014, InfoQ: What are self-organising teams? Available at http://www.infoq.com/articles/what-are-self-organising-teams (accessed Oct 27, 2014).
3. Kaltenecker, S., 2013, Platform for agile management: Evolutionary change management with Kanban. Available at http://p-a-m.org/2013/11/evolutionary-change-and-leadership-with-kanban/ (accessed Oct 27, 2014).
4. Leopold, K., 2014, klausleopold.com: Kanban and its flight levels. Available at http://www.klausleopold.com/kanban-flight-levels (accessed Oct 27, 2014).
5. Leopold, K., 2014, klausleopold.com: Why Kanban flight levels? Available at http://www.klausleopold.com/2014/01/why-kanban-flight-levels.html (accessed Oct 27, 2014).

CHAPTER 1

1. Ortmann, G., *Management in der Hypermoderne: Kontingenz und Entscheidung*, VS Verlag für Sozialwissenschaften, Wiesbaden, 2009, p. 133.
2. Anderson, D. J., *Kanban. Successful Evolutionary Change for Your Technology Business*, Blue Hole Press, Washington, 2010.

Kanban Change Leadership: Creating a Culture of Continuous Improvement,
First Edition. Klaus Leopold and Siegfried Kaltenecker.

CHAPTER 2

1. Zäpfel, G., *Grundzüge des Produktions- und Logistikmanagement*, Oldenbourg Wissenschaftsverlag, Berlin, New York, 1996.
2. Drucker, P. F., The New Productivity Challenge, *Harvard Business Review*, November–December, **6**, 1991, pp. 69–79.
3. Willke, H., Mingers, S., Piel, K., Hermsen, T., Köhler, J., Strulik, T., Vopel, O., *Wissensarbeit in intelligenten Organisationen*, Forschung an der Universität Bielefeld, Bielefeld, 18/1988, pp. 20–24.
4. Deming, W. E., *Out of the Crisis*, The Mit Press, Cambridge, 2000.
5. Ohno, T., Bodek, N., *Toyota Production System: Beyond Large-Scale Production*, Productivity Press, Cambridge, 1988.
6. Luhmann, N., *Soziale Systeme*, Suhrkamp Verlag, Frankfurt/Main, 1984.
7. Luhmann, N., *Organisation und Entscheidung*, VS Verlag für Sozialwissenschaften, Wiesbaden, 2000.
8. Anderson, D. J., *Kanban. Successful Evolutionary Change for Your Technology Business*, Blue Hole Press, Washington, 2010.

CHAPTER 4

1. Little, J. D. C., Graves, S. C., Little's Law. in: Chhajed, D., Lowe T. J. (Eds.): *Building Intuition. Insights from Basic Operations Management Models and Principles*, Springer Science + Business Media, New York, 2008.
2. Goldratt, E., *What Is This Thing Called Theory of Constraints*, North River Press, Great Barrington, 1990.

CHAPTER 7

1. Sobel, D., *Longitude. The True Story of a Lone Genius Who Solved the Greatest Scientific Problem of His Time*, Walker & Company, London, 1996.
2. Ackoff, R. L., *Management in Small Doses*, John Wiley & Sons, Inc., Hoboken, 1986.
3. Brooks, F., *The Mythical Man-Month*, Addison-Wesley, Reading, 1995.

CHAPTER 8

1. Baecker, D., *Organisation und Störung*, Suhrkamp, Frankfurt/Main, 2011, p. 28.
2. Schein, E. H., *Corporate Culture Survival Guide*, John Wiley & Sons, Inc., San Francisco, 1999.
3. Drucker, P. F., The New Productivity Challenge, *Harvard Business Review*, November–December, **6**, 1991, p. 171.
4. Senge, P., *The Fifth Discipline: The Art and Practice of the Learning Organization*, Crown Business, New York, 1990.

5. Ortmann, G., *Management in der Hypermoderne: Kontingenz und Entscheidung*, VS Verlag für Sozialwissenschaften, Wiesbaden, 2009, p. 152.
6. Dörner, D., *The Logic of Failure: Recognizing and Avoiding Error in Complex Situations*, Metropolitan, New York, 1996.
7. Twain, M., 1870, The watch. Available at http://www.readbookonline.net/readOnLine/1565/ (accessed Oct 27, 2014).

CHAPTER 9

1. Lewin, K., Frontiers in group dynamics. in: Cartwright, D. (Ed.): *Field Theory in Social Science: Selected Theoretical Papers by Kurt Lewin*, Harper & Row, New York, 1951, pp. 188–237.
2. Königswieser, R., Hillebrand, M., *Systemic Consultancy in Organisations: Concepts, Tools, Innovations*, Carl Auer, Heidelberg, 2005.
3. von Foerster, H., *Understanding Understanding: Essays on Cybernetics and Cognition*, Springer, New York, 2002.
4. Königswieser, R., Exner, A., *Systemische Intervention. Architekturen und Designs für Berater und Veränderungsmanager*, Klett-Cotta, Stuttgart, 1998, p. 17.
5. Simon, F. B., *Die Kunst nicht zu lernen*, Carl-Auer Verlag, Heidelberg, 1997, p. 14.
6. Luhmann, N., *Theory of Society*, Stanford University Press, Stanford, 2012.
7. Heitger, B., Doujak, A., *Managing Cuts and New Growth. An Innovative Approach to Change Management*, Goldegg, Vienna, 2009.

CHAPTER 10

1. Doppler K., Lauterburg, C., *Change Management. Den Unternehmenswandel gestalten*, Campus, Frankfurt/Main, 1989.
2. Jarrett, M., *Changeability: Why Some Companies Are Ready for Change—and Others Aren't*, Pearson, Harlow, 2007.
3. Duck, J. D., *The Change Monster: The Human Forces That Fuel or Foil Corporate Transformation and Change*, Three Rivers Press, New York, 2001.
4. Kübler-Ross, E., *On Death and Dying*, Scribner, New York, 1997.
5. Weick, K., *Sensemaking in Organizations*, Sage, Thousand Oaks, 1995.

CHAPTER 11

1. Ciompi, L., *Die emotionalen Grundlagen des Denkens. Entwurf einer fraktalen Affektlogik*, Vandenhoeck & Ruprecht, Göttingen, 2007.
2. Heitger, B., Doujak, A., *Managing Cuts and New Growth. An Innovative Approach to Change Management*, Goldegg, Vienna, 2009.
3. Schein, E. H., *Organizational Culture and Leadership*, Jossey-Bass, San Francisco, 2004, p. 123.
4. Hamel, G., *Leading the Revolution*, Plume, New York, 2002, p. 55.

CHAPTER 12

1. Schein, E. H., *Organizational Culture and Leadership*, Jossey-Bass, San Francisco, 2004, p. 123.
2. Senge, P., *The Fifth Discipline: The Art and Practice of the Learning Organization*, Crown Business, New York, 1990.
3. Ortmann, G., *Management in der Hypermoderne: Kontingenz und Entscheidung*, VS Verlag für Sozialwissenschaften, Wiesbaden, 2009.
4. Neuberger, O., *Führen und führen lassen. Ansätze, Ergebnisse und Kritik der Führungsforschung*, UTB, Stuttgart, 2002, p. 712.

CHAPTER 13

1. Wimmer, R., *Organisation und Beratung: Systemtheoretische Perspektiven für die Praxis*, Carl Auer, Heidelberg, 2004.
2. Ortmann, G., *Kunst des Entscheidens: Ein Quantum Trost für Zweifler und Zauderer*, Velbrück Wissenschaft, Weilerswist, 2011, p. 27.
3. Fayol, H., *Administration Industrielle et Génerale*, Guillaume, Paris, 1916.
4. Hamel, G., *The Future of Management*, Harvard Business Review Press, Boston, 2007.
5. Mintzberg, H., *Managing*, Berrett-Koehler, San Francisco, 2011, p. 3.
6. Kaltenecker, S., Ally, M., Blunden, N., Groenewald, M., Reyneke, C., 2001, Platform for agile management: High performing teams. Available at http://p-a-m.org/2011/09/high-performing-teams/ (accessed Oct 27, 2014).
7. Kaltenecker, S., Spielhofer, T., Eybl, S., Schober, J., Jäger, S., 2011, Platform for agile management: Exec summary of the study on "successful agile leadership". Available at http://p-a-m.org/2011/06/exec-summary-of-the-study-on-successful-agile-leadership/ (accessed Oct 27, 2014).
8. Katzenbach, J., Smith, D., *The Wisdom of Teams. Creating the High-Performing Organization*, Harvard Business Press, New York, 1993.
9. Raelin, J. A., *Creating Leaderful Organisations: How to Bring Out Leadership in Everyone*, Berrett-Koehler, San Francisco, 2003.
10. Wimmer, R., *Organisation und Beratung: Systemtheoretische Perspektiven für die Praxis*, Carl Auer, Heidelberg, 2004, p. 172.
11. Doppler, K., Lauterburg, C., *Change Management. Den Unternehmenswandel gestalten*, Campus, Frankfurt/Main, 1989, p. 212.
12. Weick, K., Sutcliffe, K., *Managing the Unexpected: Assuring High Performance in an Age of Complexity*, Jossey-Bass, San Francisco, 2008.
13. Denning, S., *The Leader's Guide to Radical Management: Reinventing the Workplace for the 21st Century*, Jossey-Bass, San Francisco, 2010.
14. Watzlawick, P., *How Real Is Real?: Confusion, Disinformation, Communication*, Random House, New York, 1977, p. 12.
15. Mintzberg, H., *Managing*, Berrett-Koehler, San Francisco, 2011, p. 160.
16. Isaacs, W., *Dialogue and the Art of Thinking Together*, Random House, New York, 1999.
17. Kahan, S., *Getting Change Right: How Leaders Transform Organizations from the Inside Out*, Jossey-Bass, San Francisco, 2010.